Contents

Codes and Cryptography

Dominic Welsh

Merton College and the Mathematical Institute,
University of Oxford

CLARENDON PRESS · OXFORD

OXFORD

UNIVERSITY PRESS

Great Clarendon Street, Oxford OX2 6DP

Oxford University Press is a department of the University of Oxford.
It furthers the University's objective of excellence in research, scholarship,
and education by publishing worldwide in

Oxford New York

Auckland Bangkok Buenos Aires Cape Town Chennai
Dar es Salaam Delhi Hong Kong Istanbul Karachi Kolkata
Kuala Lumpur Madrid Melbourne Mexico City Mumbai
Nairobi São Paulo Singapore Taipei Tokyo Toronto

Oxford is a registered trade mark of Oxford University Press
in the UK and in certain other countries

Published in the United States
by Oxford University Press Inc., New York

© Dominic Welsh, 1988

First published 1988
Reprinted 1989 (with corrections), 1990, 1993,
1995, 1996, 1998 (twice), 2000, 2004

A catalogue record for this book is available from the British Library

Library of Congress Cataloging in Publication Data
Welsh, Dominic.
Codes and cryptography.
Bibliography: p.
Includes index.
1. Ciphers. 2. Cryptography. I. Title.
Z103.W46 1988 652'.8 87-31354
ISBN 0 19 853287 3 (Pbk)

Printed by Antony Rowe Ltd, Chippenham, Wiltshire

Preface

This text is based on a course I have given to undergraduates at Oxford on the mathematics of Communication Theory. Its aim is to introduce the subject in as short a space as possible to students with no previous knowledge of the area.

The foundations of this book are the seminal papers by Claude Shannon on information theory and secrecy systems. The concept of entropy is fundamental to the three main problems of how to encode information (a) economically, (b) reliably, and (c) so as to preserve privacy. However, whereas in coding theory the object is to resurrect a message from an accidentally noisy environment, in cryptography the noise is deliberately superimposed so as to make it difficult for an enemy to recover information contained in the message.

Throughout the text the emphasis is on explaining the main ideas rather than proving results in complete generality. For example, most students seem to find Shannon's noisy coding theorem difficult and I have dealt only with the simplest case of binary alphabets, memoryless channels and sources. In a similar vein Chapters 4 and 9 are crash courses in algebraic coding theory and computational complexity respectively. Both of these are huge areas, rich in exciting problems, and I hope that this treatment will at least enable the reader to make a more informed choice of future options.

The mathematical prerequisites have been kept to a minimum. However, since it is impossible to understand information theory without a working knowledge of very basic probability, this has been taken for granted. Knowledge of elementary modern algebra is also assumed.

The exercises at the end of each section in the book are meant to be elementary and are to be used as a check on the understanding of the preceding principles. The problems at the end of each chapter tend to be harder. Occasionally I have used the device of giving as a problem a result I regard as interesting, together with an original reference; some of these are difficult and to be regarded more as sources of information.

The numbering system used is fairly standard. For example,

Theorem 3.4.1 refers to the first theorem of the fourth section of Chapter 3; when the reference is in the current chapter, this is shortened to Theorem 4.1.

Finally, it gives me great pleasure to thank B. J. Birch, T. F. R. G. Braun, P. J. Cameron, A. Chin, D. A. Cohen, M. J. Collins, P. J. Donnelly, D. R. Heath-Brown, F. C. Piper, J. F. Pratt, D. R. Stirzaker, and I. White for their constructive comments and suggestions about various points arising in the text. I owe a special debt to Keith Edwards, Colin McDiarmid, and Kenneth Regan, who used earlier versions of different parts of this text when giving lectures and classes associated with this course. Their suggestions have been particularly helpful. I should also like to acknowledge the co-operation and technical advice of the staff of Oxford University Press and thank Brenda Willoughby of the Mathematical Institute for her cheerful and accurate typing of what was often hieroglyphic handwriting.

Above all, I am grateful for the help given by my wife Bridget. Her criticisms and suggestions throughout the time of writing have had a very great influence on the final outcome.

D. J. A. W.

Oxford
December 1987

Notation

A few of the frequently used items which may not be familiar to all readers are listed below.

$\lceil x \rceil$ least integer greater than or equal to x

$\lfloor x \rfloor$ maximum integer not greater than x

$\ln(x)$ $\log_e x$

$\log(x)$ $\log_2 x$ (This is a departure from usual mathematical usage but is a widespread convention in this area.)

$\mathbb{Z}$ integers

$\mathbb{Z}^+$ non-negative integers

$\mathbb{Z}_n$ non-negative integers less than n

$\mathbb{Z}_n^*$ positive integers less than and coprime with n

V_n set of binary n-tuples $(x_1, \ldots, x_n)$, $x_i = 0$ or 1

Note We often write an ordered n-tuple $(x_1, \ldots, x_n)$ as $x_1 x_2 \cdots x_n$ particularly when the x_i are specified numeric entries.

$P(A)$ probability of the event A

$P(A \mid B)$ probability of the event A conditional on the event B

$O(n^k)$ a function f such that for all sufficiently large n, $|f(n)| \leq cn^k$ for some constant c

Alphabets and strings

Σ will denote an *alphabet* consisting of a finite set of *symbols* or *letters*. An ordered sequence of symbols from Σ is called a *string* or *word*. If $x = x_1 x_2 \cdots x_m$ is such a string its *length* is defined to be m and is denoted by

$$|x| = |x_1 x_2 \cdots x_m| = m.$$

If $x = x_1 \cdots x_m$ and $y = y_1 \cdots y_n$ are two strings their *concatenation* is the string $x_1 \cdots x_m y_1 \cdots y_n$.

$\Sigma^{(n)}$ the collection of strings from Σ which have length n

Σ^* the collection of all finite strings from Σ

1

Entropy = uncertainty = information

1.1 Uncertainty

Consider the following propositions.

[A]   A race between two equally matched horses is less uncertain than a race between six evenly matched horses.

[B]   The outcome of a spin on a roulette wheel is more uncertain than the throw of a die.

[C]   The throw of a fair die is more uncertain than the throw of a biased die in which the probabilities are $\frac{1}{10}$ of getting each of the numbers 1 to 5 and probability $\frac{1}{2}$ of getting a 6.

I suspect (indeed hope) that most readers will agree with each of the propositions [A], [B], and [C]. At the same time, I suspect they will find it difficult to give a formal definition of what they mean by uncertainty. One of Shannon's many achievements was to formalize this abstract idea. Suppose that X and Y are distinct random variables but that

$$P(X = 0) = p, \qquad P(X = 1) = 1 - p,$$

while

$$P(Y = 100) = p, \qquad P(Y = 200) = 1 - p,$$

where $0 < p < 1$.

We assert that any definition of uncertainty should give X and Y the same uncertainty. In other words, the uncertainty of X (and Y) should be a function *only* of the probability p. This property of the uncertainty must extend to random variables taking more than two values, and accordingly our first demand of any proposed measure of uncertainty is:

The uncertainty of a random variable Z, which takes the values a_i with probabilities p_i $(1 \leq i \leq n)$, is to be a function *only* of the probabilities $p_1, \ldots, p_n$.

Therefore we denote such a function as $H(p_1, \ldots, p_n)$ and demand of it the following properties which in view of the preceding discussion and examples we regard as minimal requirements.

(A1) $H(p_1, \ldots, p_n)$ is a maximum when $p_1 = p_2 = \ldots = p_n = 1/n$.

(A2)   For any permutation π of $(1, 2, \ldots, n)$, we have

$$H(p_1, \ldots, p_n) = H(p_{\pi(1)}, \ldots, p_{\pi(n)}).$$

In other words H must be a symmetric function of the arguments $p_1, \ldots, p_n$.
[Only the probabilities matter, not their order.]

(A3) $H(p_1, \ldots, p_n) \geq 0$ and equals zero only when one of the p_i equals 1.
[Uncertainty is an inherently positive quantity and is zero only when there is no randomness present.]

(A4) $$H(p_1, \ldots, p_n, 0) = H(p_1, \ldots, p_n).$$

[The uncertainty of a six-sided fair die is the same as a seven-sided die that has no chance of showing 7 but is otherwise fair.]

(A5) $$H\left(\frac{1}{n}, \frac{1}{n}, \ldots, \frac{1}{n}\right) \leq H\left(\frac{1}{n+1}, \frac{1}{n+1}, \ldots, \frac{1}{n+1}\right).$$

[A two-horse race is less uncertain than a three-horse race.]

(A6) $H(p_1, \ldots, p_n)$ should be a continuous function of its arguments.
[This is very natural; a small change in the probabilities should not drastically affect the uncertainty.]

(A7)   If m and n are positive integers then

$$H\left(\frac{1}{mn}, \frac{1}{mn}, \ldots, \frac{1}{mn}\right) = H\left(\frac{1}{m}, \ldots, \frac{1}{m}\right) + H\left(\frac{1}{n}, \ldots, \frac{1}{n}\right).$$

[This is the linearity condition which essentially says that the uncertainty involved in throwing an m-sided die followed by an n-sided die should be the sum of the individual uncertainties.]

Finally we lay down our last condition. It is not so intuitively obvious and needs a little thought.

(A8)   Let $p = p_1 + \ldots + p_m$ and $q = q_1 + \ldots + q_n$ where each p_i and q_j are non-negative; then, if p and q are positive, with $p + q = 1$, we must have

$$H(p_1, \ldots, p_m, q_1, \ldots, q_n) = H(p, q) + pH(p_1/p, \ldots, p_m/p)$$
$$+ qH(q_1/q, \ldots, q_n/q).$$

[Think of a race in which there are m black horses and n grey horses, with p_i the probability the ith black horse wins. The total uncertainty in the outcome is the uncertainty associated with a grey or black winner plus the weighted sum of the uncertainties given that the winner is respectively grey or black.]

From these assumptions we can prove:

Theorem 1 *Let $H(p_1, \ldots, p_n)$ be a function defined for any integer n and for all values of $p_1, \ldots, p_n$ satisfying $p_i \geq 0$ and*

$$\sum_{i=1}^{n} p_i = 1.$$

If H is to satisfy the axioms (A1)–(A8), then

(1) $$H(p_1, p_2, \ldots, p_n) = -\lambda \sum_k p_k \log p_k,$$

with λ any positive constant and where the sum is over those k for which $p_k > 0$.

Although Theorem 1 is a beautiful and motivating result its proof is not an integral part of this course and is deferred to Appendix I. It does however add plausibility to the following definitions.

Given a random variable X that takes a finite set of values with probabilities $p_1, p_2, \ldots, p_n$, we define the *uncertainty* or *entropy* of X to be

(2) $$H(X) = -\sum_k p_k \log_2 p_k$$

where we are taking logarithms to the base 2 (for historical reasons) and where again the sum is over those k with $p_k > 0$.

Note 1 The sum on the right-hand side of (2) is going to occur frequently in what follows and, in order to avoid repeating a caveat about the probabilities p_k being strictly positive, we will henceforth *assume* that, in any sum of this type, the probabilities are nonzero.

Note 2 The conditions (A1)–(A8) are essentially the axioms for entropy proposed by Shannon (1948). They are not minimal and there exists an extensive literature on this subject; see Aczél and Daróczy (1975).

Exercises 1.1 1. Which race has greater uncertainty: a handicap in which there are seven runners, three having probability $\frac{1}{6}$ of winning and four having probability $\frac{1}{8}$ or a selling plate in which there are eight runners with two horses having $\frac{1}{4}$ probability of winning and six horses each having $\frac{1}{12}$ probability?

2. Verify that the entropy function defined by (2) satisfies the conditions (A1)–(A8).

1.2 Entropy and its properties

We have agreed that, for any random variable X taking only a finite number of values with probabilities $p_1, \ldots, p_n$ such that

$$\sum p_i = 1 \quad \text{and} \quad p_i > 0 \quad (1 \le i \le n),$$

we define the entropy of X by

$$H(X) = -\sum_{k=1}^{n} p_k \log p_k.$$

In an exactly analogous way, if X is a random vector which takes only a finite number of values $u_1, u_2, \ldots, u_m$ (say), we define its *entropy* by

(1)
$$H(X) = -\sum_{k=1}^{m} p(u_k) \log p(u_k).$$

For example, when X is a two-dimensional random vector, say $X = (U, V)$ with

$$p_{ij} = P(U = a_i, V = b_j),$$

then we often write

$$H(X) = H(U, V) = -\sum_{i,j} p_{ij} \log p_{ij}.$$

More generally, if $X_1, X_2, \ldots, X_m$ is any collection of random variables each taking only a finite set of values, then we may regard $X = (X_1, \ldots, X_m)$ as a random vector taking only a finite set of values and define the *joint entropy* of $X_1, \ldots, X_m$ by

(2)
$$H(X_1, \ldots, X_m) = H(X)$$

$$= -\sum p(x_1, \ldots, x_m) \log p(x_1, \ldots, x_m)$$

where $p(x_1, \ldots, x_m) = P(X_1 = x_1, X_2 = x_2, \ldots, X_m = x_m)$. It is easy to prove that:

(3) $H(X) = 0$ if and only if X is a constant.

An upper bound for H is given by the following:

Theorem 1

For any n,

$$H(p_1, \ldots, p_m) \leq \log_2 n$$

with equality if and only if $p_1 = p_2 = \ldots = p_n = n^{-1}$.

Proof

Write

$$\log_2 x = \log_2 e \log_e x.$$

Since the logarithm is a convex function (that is its curve always lies below its tangent) we have

$$\log_e x \leq x - 1$$

with equality if and only if $x = 1$. Hence, if $(q_1, \ldots, q_n)$ is any probability vector, then

$$\log_e(q_k/p_k) \leq (q_k/p_k) - 1,$$

with equality if and only if $q_k = p_k$. Hence,

$$\sum p_i \log_e(q_i/p_i) \leq \sum q_k - \sum p_k = 0,$$

which implies

$$\sum p_i \log q_i \leq \sum p_i \log p_i.$$

Putting $q_i = 1/n$ gives

$$H(p_1, \ldots, p_n) = -\sum p_i \log_2 p_i \leq \log_2 n$$

with equality as stated. ☐

In the above proof we have proved a very useful inequality, which we name and state as follows.

Key Lemma

If $(p_i : 1 \leq i \leq n)$ is a given probability distribution, then the minimum of

$$G(q_1, \ldots, q_n) = -\sum p_i \log q_i$$

over all probability distributions $(q_1, \ldots, q_n)$ is achieved when $q_k = p_k$ $(1 \leq k \leq n)$.

Note

Here and elsewhere we mean by a probability distribution

$(p_1, \ldots, p_n)$, any set of nonnegative numbers p_i such that

$$\sum p_i = 1.$$

Using this lemma, it is straightforward to prove the following.

Theorem *If X and Y are any two random variables taking only finitely many*
2 *values, then*

$$H(X, Y) \le H(X) + H(Y),$$

with equality holding only when X and Y are independent.

Proof Suppose that

$$r_i = P(X = a_i) \quad (1 \le i \le m), \qquad s_j = P(Y = b_j) \quad (1 \le j \le n),$$
$$t_{ij} = P(X = a_i, Y = b_j) \quad (1 \le i \le m, \ 1 \le j \le n).$$

Then

$$H(X) + H(Y) = -\left(\sum_i r_i \log r_i + \sum_j s_j \log s_j\right)$$

$$= -\left(\sum_i \sum_j t_{ij} \log r_i + \sum_j \sum_i t_{ij} \log s_j\right)$$

because

$$r_i = \sum_j t_{ij}, \qquad s_j = \sum_i t_{ij}.$$

Hence

$$H(X) + H(Y) = -\sum_i \sum_j t_{ij} \log (r_i s_j)$$

$$\ge -\sum \sum t_{ij} \log t_{ij} = H(X, Y),$$

by the preceding Key Lemma. Equality will hold only when

$$r_i s_j = t_{ij} \quad (1 \le i \le m, \ 1 \le j \le n).$$

But this is exactly the condition that X and Y are independent.   □

By a straightforward extension of this method, one can prove:

(4) $$H(X_1, \ldots, X_n) \le H(X_1) + \ldots + H(X_n),$$

with equality holding only when $X_1, \ldots, X_n$ are mutually independent;

(5)
$$H(U, V) \leq H(U) + H(V)$$

for any pair of random vectors (U, V) with equality holding only when U and V are independent random vectors.

The proofs (which follow in exactly the same way as does that of Theorem 2 from the Key Lemma) are left to the reader.

Exercises 1.2

1. Two fair dice are thrown; X denotes the value obtained by the first, and Y denotes the value obtained by the second. Verify that $H(X, Y) = H(X) + H(Y)$. Show that, if $Z = X + Y$, then

$$H(Z) < H(X, Y).$$

2. Show that, for any random variable X,

$$H(X, X^2) = H(X).$$

3. Show that, for any sequence of random variables $(X_i : 1 \leq i < \infty)$,

$$H(X_1, \ldots, X_n) \leq H(X_1, \ldots, X_{n+1}).$$

1.3 Conditional entropy

Suppose that X is a random variable on a probability space Ω and A is an event in Ω. If X takes a finite set of values $\{a_i : 1 \leq i \leq m\}$, it is natural to define the *conditional entropy* of X given A by

$$H(X \mid A) = - \sum_{k=1}^{m} P(X = a_k \mid A) \log P(X = a_k \mid A).$$

In the same way, if Y is any other random variable taking values b_k $(1 \leq k \leq m)$, we define the *conditional entropy* of X given Y by

$$H(X \mid Y) = \sum_{j} H(X \mid Y = b_j) P(Y = b_j).$$

We think of $H(X \mid Y)$ as the uncertainty of X given a particular value of Y, averaged over the range of values that Y can take.

Fairly trivial consequences of the definitions are:

(1)
$$H(X \mid X) = 0,$$

(2)
$$H(X \mid Y) = H(X) \quad \text{if } X \text{ and } Y \text{ are independent.}$$

Example   Let X be the value obtained by throwing a fair die. Let Y be another random variable determined by the same experiment with Y equalling 1 if the value thrown is odd and 0 otherwise. Since the die is fair

$$H(X) = \log 6,$$
$$H(Y) = \log 2,$$

and

$$H(X \mid Y) = \log 3. \qquad \Box$$

When U and V are random vectors, we naturally extend the definition of conditional entropy by defining

(3) $$H(U \mid V) = \sum_j H(U \mid V = v_j)P(V = v_j),$$

where the sum, as usual, is over the (finite) range of values v_j that V has a positive probability of taking.

As a first example of the way in which $H(U \mid V)$ measures the uncertainty about U contained in V we prove:

(4) $$H(U \mid V) = 0 \text{ if and only if } U = g(V) \text{ for some function } g.$$

Proof   The right-hand side of (3) is the sum of a finite number of nonnegative quantities. Hence, for it to be zero, we need $H(U \mid V = v_j)$ to be zero for each j. But again each of these nonnegative quantities is zero only if U is uniquely determined by V. $\qquad \Box$

Slightly more care gives the following result, which expresses mathematically the idea that our definition of conditional entropy of X given Y correctly measures the remaining uncertainty.

Theorem 1   *For any pair of random variables X and Y that take only a finite set of values,*

$$H(X, Y) = H(Y) + H(X \mid Y).$$

Proof   Without loss of generality, we suppose X and Y take integer values and, where necessary, we let $p_{ij} = P(X = i, Y = j)$. Now

$$H(X, Y) = -\sum_i \sum_j P(X = i, Y = j)\log P(X = i, Y = j)$$

$$= -\sum_i \sum_j P(X = i, Y = j)\log P(X = i \mid Y = j)P(Y = j)$$

$$= -\sum \sum p_{ij} \log P(X = i \mid Y = j) - \sum \sum p_{ij} \log P(Y = j)$$

$$= -\sum\sum P(X=i\,|\,Y=j)P(Y=j)\log P(X=i\,|\,Y=j) + \mathrm{H}(Y)$$

$$= -\sum_j P(Y=j)\sum_i P(X=i\,|\,Y=j)\log P(X=i\,|\,Y=j) + \mathrm{H}(Y)$$

$$= \sum_j P(Y=j)\mathrm{H}(X\,|\,Y=j) + \mathrm{H}(Y)$$

$$= \mathrm{H}(Y) + \mathrm{H}(X\,|\,Y)\quad\text{as required.}\qquad\square$$

The method of proving the above theorem, (essentially definition chasing) carries over to the case of several variables. More precisely, we can prove the following.

Theorem 2   *If U and V random vectors each taking only a finite set of values, then*

$$\mathrm{H}(U,\,V) = \mathrm{H}(V) + \mathrm{H}(U\,|\,V).$$

Proof   Follow through the proof of Theorem 1, but instead of X and Y taking integer values i and j, we have U and V taking values u_i and v_j, where u_i and v_j are prescribed vectors. $\square$

The following result is an immediate consequence.

Corollary   *For any pair of random vectors X and Y, $\mathrm{H}(X\,|\,Y) \le \mathrm{H}(X)$, with equality if and only if X and Y are independent.*

Proof

$$\mathrm{H}(X\,|\,Y) = \mathrm{H}(X,\,Y) - \mathrm{H}(Y)$$

But $\mathrm{H}(X,\,Y) \le \mathrm{H}(X) + \mathrm{H}(Y)$, with equality if and only if X and Y are independent, and the result follows. $\square$

Exercises 1.3

1. Show that, for any random variable X,
$$\mathrm{H}(X^2\,|\,X) = 0,$$
but give an example to show that $\mathrm{H}(X\,|\,X^2)$ is not always zero.

2. The random variable X takes the integer values $1, 2, \ldots, 2N$ with equal probability. The random variable Y is defined by $Y = 0$, if X is even, but $Y = 1$ if X is odd. Show that
$$\mathrm{H}(X\,|\,Y) = \mathrm{H}(X) - 1$$
but that $\mathrm{H}(Y\,|\,X) = 0$.

1.4 Information

R. V. L. Hartley in 1928 seems to have been the first to attempt to assign a quantitative measure to the concept of information. The rationale behind this attempt can be partly explained as follows.

Suppose E_1 and E_2 are two events in a probability space Ω associated with some experiment and suppose that the function I is to be our measure of information. If E_1 and E_2 have probabilities p_1 and p_2 respectively, then it could be argued that any natural measure of the information content should satisfy

(1)
$$I(p_1 p_2) = I(p_1) + I(p_2)$$

on the grounds that, for two independent realizations of the experiment, the information that the results of these experiments turned out to be E_1 followed by E_2 should be the sum of the information obtained by carrying out the experiments separately.

Granting that (1) has some validity, and wishing to make our measure of information non-negative and continuous in p, both natural assumptions, we are left with little alternative but to *define* the *information* I of an event E of positive probability by

(2)
$$I(E) = -\log_2 P(E)$$

where we have chosen the base 2 for our logarithms in order to conform with modern conventions. (Hartley originally used logarithms to the base 10.)

Example Suppose we have a source which emits a string of binary digits 0 and 1, each with equal probability and independently for successive digits. Let E be the event that the first n digits are alternately zeros and ones. Then clearly

$$I(E) = -\log_2(1/2^n) = n$$

and the same applies to any prescribed n-sequences of digits. □

Thus the 'information-theoretic' unit of information, namely the *bit*, corresponds naturally to the use of the word 'bit' to mean a binary digit in present-day computing terminology.

We extend this concept of information to cover random variables and vectors as follows. Suppose U is a random vector taking the values $u_1, \ldots, u_m$ with probabilities $p_1, \ldots, p_m$ respectively. Then each of the elementary events $\{U = u_k\}$ $(1 \le k \le m)$ has an associated information equal to $-\log_2 p_k$, and we notice that the entropy of the vector U is given by

$$H(U) = -\sum p_k \log_2 p_k = \sum p_k I(\{U = u_k\}),$$

so that $H(U)$ has a natural interpretation as the mean value of the information associated with the elementary events determined by U.

More generally, if U and V are any two random vectors, we define the *information about U conveyed* by V to be the quantity

$$I(U \mid V) = H(U) - H(U \mid V).$$

In other words, $I(U \mid V)$ measures the amount of uncertainty about U that is removed by V.

Trivially, we see that

(3) $$I(U \mid U) = H(U),$$

(4) $I(U \mid V) = 0$ if and only if U and V are independent.

Proof   The result follows immediately from the earlier remark that $H(U) = H(U \mid V)$ only when U and V are independent.   □

A somewhat surprising symmetry in I is the following result which, as far as I can see, has no intuitive explanation.

$$I(U \mid V) = H(U) - H(U \mid V)$$

(5)
$$= H(U) - [H(U, V) - H(V)]$$
$$= H(U) + H(V) - H(U, V)$$
$$= I(V \mid U).$$

Exercises   1. Which has got the greater information content: a sequence of 10 letters or
1.4   a sequence of 26 digits from the set $\{0, 1, \dots, 9\}$? [Assume all sequences
                   are equiprobable.]
           2. A fair die is thrown. Show that the information about the value of the die
                   given by the knowledge that it is prime is given by $\log_2 \frac{3}{2}$.

1.5 Conclusion

To sum up, we have shown that, essentially, uncertainty and information are the same quantities, the removal of uncertainty being equated with the giving of information. Both are measured by the mathematical concept of entropy, which is uniquely defined (up to a multiplicative constant) by the quantity

$$H = -\sum p_i \log_2 p_i.$$

Convention demands that logarithms are taken to the base 2, in which case the unit of entropy is a *bit*.

PROBLEMS 1

1. A disc jockey has a vocabulary of 10 000 words, and he utters 1000 at random (repetitions allowed). Show that the information content of his 1000 words is much less than that of a TV picture of 500 rows and 600 columns with each pixel taking one of 16 brightness levels.

2. If X and Y are discrete random variables taking only a finite number of values, show that

$$H(X + Y \mid X) = H(Y \mid X).$$

Show that

$$H(g(X, Y) \mid X) = H(Y \mid X).$$

does not hold generally for $g : \mathbb{R}^2 \to \mathbb{R}$.

3. If $X_1, X_2, \ldots$ is any sequence of random variables, and Y is any other random variable, prove that

$$H(X_1, \ldots, X_n \mid Y) \le H(X_1, \ldots, X_{n+1} \mid Y)$$

for any positive integer n.

4. A statistical survey of married couples shows that 70% of men have dark hair, that 25% of girls are blonde, and that 80% of blonde girls marry dark-haired men. How much information about the colour of a man's hair is conveyed by the colour of his wife's hair?

5. If X, Y, Z are random vectors, each taking only finitely many values, prove that

$$H(Y \mid X) + H(Z \mid X) \ge H(Y, Z \mid X).$$

6. Show that, for any random vector Y and any set of random variables $X_1, \ldots, X_{n+1}$,

$$H(Y \mid X_1, \ldots, X_n) \ge H(Y \mid X_1, \ldots, X_{n+1}).$$

7. If X and Y are two random variables and f and g are any two functions, prove that

$$H(f(X), g(Y)) \le H(X, Y).$$

8. A random variable X has the binomial distribution with parameters n

and p; that is, for $0 \le k \le n$,

$$P(X = k) = \binom{n}{k} p^k q^{n-k},$$

where $0 < p < 1$ and $q = 1 - p$.
Show that

$$H(X) \le -n(p \log p + q \log q).$$

9. A random variable X has the geometric distribution, taking integer values $k = 0, 1, 2, \ldots$ with

$$p_k = P(X = k) = pq^k,$$

where $p > 0$ and $q + p = 1$. Show that, if we extend the notion of entropy to define

$$H(X) = - \sum_{k=0}^{\infty} p_k \log p_k$$

whenever the right hand series converges, then, in this particular case,

$$H(X) = -(p \log p + q \log q)/p.$$

10. Call two random variables X and Y *equivalent* if $H(X \mid Y) = 0$ and $H(Y \mid X) = 0$. Show that, if X and Y are equivalent and Y and Z are equivalent, then X and Z are equivalent.

11. Define the *distance* between two random variables X and Y by

$$d(X, Y) = H(X \mid Y) + H(Y \mid X).$$

Show that for any three random variables X, Y, Z,

$$d(X, Y) + d(Y, Z) \ge d(X, Z).$$

12. Suppose that X is a random variable taking values $v_1, \ldots, v_n$. Show that, if $E(X) = \mu$ and X is a random variable of maximum entropy subject to these constraints, then

$$p_j = P(X = v_j) = A e^{-\alpha v_j},$$

where A and α are constants determined by $E(X) = \mu$ and $\sum p_j = 1$.
Note: the above example is an illustration of the *principle of maximum entropy*: this is an extension of Laplace's principle of insufficient reason; it is much used in statistical mechanics, image processing, and the like as a principle for choosing an a priori distribution subject to various constraints. See for example Guiasu and Shenitzer (1985).

2

The noiseless coding theorem
for memoryless sources

2.1 Memoryless sources

In this chapter we prove the first and easier of Shannon's two main theorems, for the simplest class of sources.

The Concise Oxford Dictionary defines a *source* as 'spring, fountain head, from which stream issues'. In its complete generality, this is how it is regarded in information theory, although typically we will be regarding a source as a stream of symbols from some finite alphabet. The source usually has some random mechanism which is based on the statistics of the situation being modelled. This random mechanism can be pretty complicated, but—for the moment—we will concentrate on the following very special and simple case. If X_i denotes the ith symbol produced by the source, then we stipulate that, for each symbol a_j, the probability

$$P(X_i = a_j) = p_j$$

is independent of i and also is independent of all previous or future symbols emitted. In other words, $X_1, X_2, \ldots$ is just a sequence of identically distributed, independent random variables. Such a source is called a *zero-memory* or *memoryless source* and its entropy H is defined by

$$H = -\sum p_j \log p_j$$

where the sum is over the set of j such that $p_j > 0$.

Exercise 2.1

1. If $\mathscr{S}$ is any zero-memory source with alphabet Σ, the *n-th order extension* of $\mathscr{S}$ is a zero-memory source $\mathscr{S}^{(n)}$, with alphabet the set $\Sigma^{(n)}$ consisting of all strings of length n of symbols from Σ, and such that the probability of any particular string σ is given by the probability that it is the string of the first n symbols emitted by $\mathscr{S}$. Show that $\mathscr{S}^{(n)}$ has entropy given by

$$H(\mathscr{S}^{(n)}) = nH(\mathscr{S}).$$

2.2   Instantaneous and uniquely decipherable codes

The main problem settled in this chapter is this. Suppose that we have a memoryless source $\mathscr{S}$ which emits symbols from an alphabet $W = \{w_1, \ldots, w_m\}$ with probabilities $\{p_1, \ldots, p_m\}$ respectively. For pedagogical reasons, we call the elements of W *source words* and ask the following question. If Σ is an alphabet of D symbols, how can we encode the source words w_i using symbols from Σ so as to achieve as economic an encoding as possible?

Example   Suppose the source $\mathscr{S}$ emits four source words a, b, c, d with probabilities

$$p_a = 0.9, \qquad p_b = 0.05, \qquad p_c = p_d = 0.025$$

Then, comparing the encodings

$$a \rightsquigarrow 0, \qquad b \rightsquigarrow 111, \qquad c \rightsquigarrow 110, \qquad d \rightsquigarrow 101,$$

and

$$a \rightsquigarrow 00, \qquad b \rightsquigarrow 01, \qquad c \rightsquigarrow 10, \qquad d \rightsquigarrow 11,$$

it is clear that the average length of an encoded source word is 1.2 in the first code, and 2 in the second. $\square$

More formally, an *encoding* or *code* is a map f from $\{w_1, \ldots, w_m\}$ into Σ^* where Σ^* denotes the collection of finite strings of symbols from Σ. A *message* is any finite string of source words and, if

$$m = w_{i_1} \ldots w_{i_k}$$

and if f is a code, then the *extension* of f to W^* is defined in the obvious way by the concatenation

$$f(m) = f(w_{i_1}) f(w_{i_2}) \ldots f(w_{i_k}).$$

A code f is *uniquely decipherable* if any finite string from Σ^* is the image of at most one message. The strings $f(w_i)$ are called the *code words*, and the integers $|f(w_i)|$ are the *word lengths* of f. The *average length* of the code f is $\langle f \rangle$ defined by

$$\langle f \rangle = \sum_{i=1}^{m} p_i \, |f(w_i)| .$$

A code f is *instantaneous* or a *prefix code* if there do not exist distinct w_i and w_j such that $f(w_i)$ is a prefix of $f(w_j)$. Here, as one would expect, we are using *prefix* in the obvious sense that, if $x, y \in \Sigma^*$, then x is a prefix of y if there exists $z \in \Sigma^*$ such that $xz = y$.

Instantaneous codes are clearly uniquely decipherable. In fact, they have the stronger property that an instantaneous code can be decoded 'on line' without looking into the future.

Example Suppose $\Sigma = \{0, 1\}$ and there are four source words $w_1, \ldots, w_4$. An instantaneous code is

$$f(w_1) = 0, \qquad f(w_2) = 10, \qquad f(w_3) = 110, \qquad f(w_4) = 1110.$$

Thus, for example, a message 01101001010010 would be decoded as $w_1\, w_3\, w_2\, w_1\, w_2\, w_2\, w_1\, w_2$. (This is an example of what is known as a *comma* code, since we are clearly using 0 as a means of signalling the end of words.) □

Not every uniquely decipherable code is instantaneous.

Example Suppose $W = \{w_1, w_2\}$, $\Sigma = \{0, 1\}$, and the code g is defined by

$$g(w_1) = 0, \qquad g(w_2) = 01.$$

This is clearly not instantaneous, but can be easily verified to be uniquely decipherable by working backwards from the end of the message. □

It is clear that uniquely decipherable codes are a much more difficult concept than instantaneous codes. Luckily, and surprisingly, we will show that we can restrict attention to instantaneous codes in our search for uniquely decipherable codes having minimum average length.

Note Although we have defined a code as a map, we often identify it with the collection $\mathcal{C}$ of codewords.

Exercise 2.2 1. Show that, for any positive integer m there exists an instantaneous code over $\{0, 1\}$ that has words of all lengths in the set $\{1, \ldots, m\}$.

2.3 The Kraft–McMillan inequalities

In this section we prove two fundamental inequalities which substantiate our earlier remark that we can essentially forget about the concept of unique decipherability and restrict attention to instantaneous codes.

First we state the inequalities.

KRAFT'S INEQUALITY

If Σ is an alphabet of size D and W contains N words, then a necessary and sufficient condition that there exists an instantaneous code $f: W \to \Sigma^*$ with word lengths $l_1, \ldots, l_N$ is that

(1)
$$\sum_{i=1}^{N} D^{-l_i} \leq 1.$$

McMILLAN'S INEQUALITY

If Σ is an alphabet of size D and W contains N words, then a necessary condition that there exists a uniquely decipherable code with code words of length $l_1, \ldots, l_N$ is that (1) holds.

Combining the two inequalities we get:

Theorem 1 *A uniquely decipherable code with prescribed word lengths exists if and only if an instantaneous code with the same word lengths exists.*

PROOF OF KRAFT'S INEQUALITY

Suppose that the set $\{l_1, \ldots, l_N\}$ satisfies

$$\sum_{i=1}^{N} D^{-l_i} \leq 1.$$

Rewrite the inequality in the form

$$\sum_{j=1}^{l} n_j D^{-j} \leq 1,$$

where n_j is the number of l_i equal to j, and $l = \max l_i$.

Rewrite the inequality again in the form

(2)
$$n_l \leq D^l - n_1 D^{l-1} - \ldots - n_{l-1} D.$$

Since the n_j are all non-negative, we successively get from (2) the inequalities

(3)
$$n_{l-1} \leq D^{l-1} - n_1 D^{l-2} - \ldots - n_{l-2} D,$$
$$n_{l-2} \leq D^{l-2} - n_1 D^{l-3} - \ldots - n_{l-3} D,$$
$$\vdots$$
$$n_3 \leq D^3 - n_1 D^2 - n_2 D,$$
$$n_2 \leq D^2 - n_1 D,$$
$$n_1 \leq D.$$

These inequalities are the key to constructing a code with the given word lengths.

We first choose n_1 words of length 1, using distinct letters from Σ. This leaves $D - n_1$ symbols unused, and we can form $(D - n_1)D$ words of length 2 by adding a letter to each of these.

Choose our n_2 words of length 2 arbitrarily from these, and this leaves $D^2 - n_1 D - n_2$ prefixes of length 2.

These can be used to form $(D^2 - n_1 D - n_2)D$ words of length 3, from which we choose n_3 arbitrarily, and so on. Continue in this way, each time preserving the property that no word is a prefix of another.

In each case, we find that the inequalities (3) enable us to make this choice. Thus we end up with an instantaneous code with the prescribed code lengths. $\qquad\square$

This proves that the numerical condition (1) is sufficient for the existence of an instantaneous code. Although Kraft also showed the necessity of condition (1), it is an immediate consequence of McMillan's inequality which we next prove. The proof given is much simpler than McMillan's original proof and is due to Karush (1961).

PROOF OF McMILLAN'S INEQUALITY

Suppose we have a uniquely decipherable code $\mathscr{C}$ with word lengths $l_1, \ldots, l_N$.

If $l = \max l_i$, then, for any positive integer r, we have

(4)
$$(D^{-l_1} + \ldots + D^{-l_N})^r = \sum_{i=1}^{rl} b_i D^{-i},$$

where b_i is a non-negative integer. But, by an argument reminiscent of the way probability generating functions are manipulated, the integer b_i justs counts the number of ways in which a string of length i of symbols from the alphabet Σ can be made up by stringing together r words of lengths chosen from the set $\{l_1, \ldots, l_N\}$.

But, if the code $\mathscr{C}$ is uniquely decipherable, it must be the case that any string of length i formed from codewords must correspond to at most one sequence of code words. Hence we must have

$$b_i \le D^i \quad (1 \le i \le rl).$$

Hence, substituting in (4), we obtain

$$\left(\sum_{k=1}^{N} D^{-l_k}\right)^r \le lr.$$

Therefore

$$\sum_{k=1}^{N} D^{-l_k} \leq l^{1/r} r^{1/r},$$

and, since r is an arbitrary positive integer, we can let $r \to \infty$ in the right-hand side to get McMillan's equality. $\square$

Exercise 2.3

1. What is the maximum number of words in a binary instantaneous code in which the maximum word length is 7?

2.4   The noiseless coding theorem for memoryless sources

Consider now the following situation. We have a memoryless source $\mathscr{S}$ which emits words $w_1, \ldots, w_m$ with probabilities $p_1, \ldots, p_m$ respectively, each word emitted being chosen independently of all other words. Our problem is: given such a source together with an alphabet Σ, find a uniquely decipherable code whose average word length is as small as possible. Such a code we describe as *compact*.

A heuristic approach to this problem might proceed as follows. The source $\mathscr{S}$ has entropy

$$H = -\sum p_i \log p_i.$$

The maximum entropy in an alphabet of D letters is $\log D$. Hence the number of symbols of the alphabet needed on the average to encode a word of the source should be about $H/\log D$.

This crude idea is now made precise.

Theorem 1

If a memoryless source has entropy H, then any uniquely decipherable code for this source into an alphabet of D symbols must have length at least $H/\log D$. Moreover, there exists such a uniquely decipherable code having average word length less than or equal to $1 + H/\log D$.

Proof

Suppose $\mathscr{C}$ is a uniquely decipherable code with word lengths $l_1, \ldots, l_N$. Suppose also that the probabilities of words corresponding to these lengths being emitted are $p_1, \ldots, p_N$ respectively. Thus,

$$H = -\sum p_i \log p_i$$

and the average length of $\mathscr{C}$ is given by

$$\ell(\mathscr{C}) = \sum p_i l_i.$$

By the Kraft–McMillan inequalities, we know that

$$G = \sum D^{-l_i} \leq 1.$$

Define q_i $(1 \leq i \leq N)$ by

$$q_i = D^{-l_i}/G,$$

so that $(q_1, \ldots, q_N)$ is a probability distribution. Apply the Key Lemma (§1.2) and we get

$$H = - \sum p_i \log p_i \leq - \sum p_i \log q_i.$$

But

$$\log q_i = -l_i \log D - \log G.$$

Hence

$$H \leq \left(\sum p_i l_i \right) \log D + \left(\sum p_i \right) \log G.$$

But $G \leq 1$ by the Kraft–McMillan inequalities, and hence, as required,

$$H \leq \ell(\mathscr{C}) \log D.$$

To prove the upper bound, we just choose our word lengths $l_1, \ldots, l_N$ by the rule that for each i, the length l_i is the minimum integer satisfying

(1) $$p_i^{-1} \leq D^{l_i}.$$

But, since $p_1 + \ldots + p_N = 1$, this implies

$$\sum_{1}^{N} D^{-l_i} \leq 1,$$

so that we know there exists a uniquely decipherable (indeed instantaneous) code with these word lengths.

But, since (1) is equivalent to

$$l_i \log D \geq -\log p_i$$

and l_i is minimal with respect to this property, we know that

$$l_i < 1 - (\log p_i)/\log D$$

Hence we know that

$$\ell(\mathscr{C}) = \sum p_i l_i < 1 + H/\log D.$$

$\square$

Exercises 2.4

1. When encoding n equiprobable source words over the binary alphabet, the noiseless coding theorem shows that the average word length $\ell(\mathscr{C})$ of any compact uniquely decipherable code satisfies

$$\log_2 n \le \ell(\mathscr{C})$$

for which values of n does equality hold?

2. Compare the noiseless coding theorem bounds with the length of a compact encoding of $2^k - 1$ equiprobable words over $\{0, 1\}$.

2.5   Constructing compact codes

Suppose that we are given a memoryless source $\mathscr{S}$ of words $w_1, \ldots, w_N$ with probabilities $p_1, \ldots, p_N$, respectively, and that we wish to find a compact code $\mathscr{C}$ for $\mathscr{S}$ over an alphabet Σ. We know from the noiseless coding theorem that this average length must satisfy the fairly tight bounds

(1)
$$\mathrm{H}(\mathscr{S})/\log D \le \ell(\mathscr{C}) \le \mathrm{H}(\mathscr{S})/\log D + 1,$$

but it is also easy to see that the lower bound can only be achieved when the p_i are certain integral powers of D. From the Kraft–McMillan inequality we have:

(2)   If there exists a compact uniquely decipherable code of average length l, then there exists a compact instantaneous code of average length l.

Hence we may restrict attention to instantaneous codes. We now describe a method devised by Huffman in 1952 for constructing a compact instantaneous code for the above source $\mathscr{S}$ in the case where Σ is a binary alphabet. We first have to prove some properties of a compact instantaneous code $\mathscr{C}$ over $\Sigma = \{0, 1\}$. We will use $\ell(w)$ to denote the length of the word w in $\mathscr{C}$.

Lemma 1   *A compact code for a source with just two words w_1 and w_2 is*

$$w_1 \to 0, \qquad w_2 \to 1.$$

Proof   Obvious.   □

Lemma 2   *If $\mathscr{C}$ is instantaneous and compact, and $p_i > p_j$, then $\ell(w_i) \le \ell(w_j)$.*

Proof   If not, form a new code $\mathscr{C}'$ from $\mathscr{C}$ by interchanging the encodings of w_i and w_j. Then the average length is reduced, and we still have an instantaneous code.   □

Lemma 3

If $\mathscr{C}$ is instantaneous and compact, then, among the code words in $\mathscr{C}$ which have maximum length, there must be at least two agreeing in all but the last digit.

Proof

Suppose not; then we can drop the last digit off all these code words of maximal length and still have an instantaneous code, contradicting $\mathscr{C}$ being compact. □

THE HUFFMAN CODING ALGORITHM

Without loss of generality, we may assume that the source $\mathscr{S}$ has its collection of source words $\{w_1, \ldots, w_N\}$ ordered so that the probabilities p_i of emitting w_i satisfy

$$p_1 \geq p_2 \geq \ldots \geq p_N.$$

The Huffman procedure constructs recursively a succession of sources $\mathscr{S}_0, \mathscr{S}_1, \ldots, \mathscr{S}_{N-2}$ such that $\mathscr{S}_0 = \mathscr{S}$ and $\mathscr{S}_k$ is obtained from $\mathscr{S}_{k-1}$ by identifying the two least probable symbols of $\mathscr{S}_{k-1}$ with a single symbol σ in $\mathscr{S}_k$. The probability that σ is emitted from $\mathscr{S}_k$ is taken to be the sum of the probabilities of its two constituent symbols in $\mathscr{S}_{k-1}$.

Thus $\mathscr{S}_1$ is obtained from $\mathscr{S}_0$ by identifying w_N and w_{N-1} into a single symbol w_{N-1} occurring with probability $p_N + p_{N-1}$. At each stage we have a source with one fewer symbol until after $N-2$ such reductions we arrive at a source $\mathscr{S}_{N-2}$ which has only two symbols. The transition between $\mathscr{S}_{j-1}$ and $\mathscr{S}_j$ is best seen diagrammatically as shown in Fig. 1.

Given an encoding $\sigma_1, \ldots, \sigma_t$ of the source $\mathscr{S}_j$ as shown,

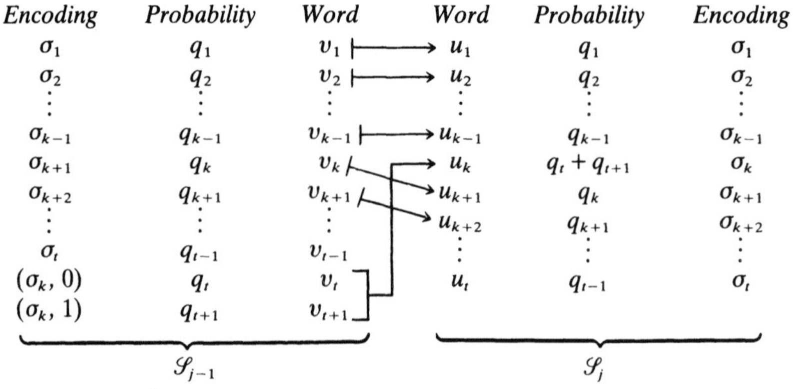

Encoding	Probability	Word	Word	Probability	Encoding
σ_1	q_1	$v_1 \longmapsto$	u_1	q_1	σ_1
σ_2	q_2	$v_2 \longmapsto$	u_2	q_2	σ_2
$\vdots$	$\vdots$	$\vdots$	$\vdots$	$\vdots$	$\vdots$
σ_{k-1}	q_{k-1}	$v_{k-1} \longmapsto$	u_{k-1}	q_{k-1}	σ_{k-1}
σ_{k+1}	q_k	v_k	u_k	$q_t + q_{t+1}$	σ_k
σ_{k+2}	q_{k+1}	v_{k+1}	u_{k+1}	q_k	σ_{k+1}
$\vdots$	$\vdots$	$\vdots$	u_{k+2}	q_{k+1}	σ_{k+2}
σ_t	q_{t-1}	v_{t-1}	$\vdots$	$\vdots$	$\vdots$
$(\sigma_k, 0)$	q_t	v_t	u_t	q_{t-1}	σ_t
$(\sigma_k, 1)$	q_{t+1}	v_{t+1}			

$$\underbrace{\hspace{7cm}}_{\mathscr{S}_{j-1}} \qquad \underbrace{\hspace{4cm}}_{\mathscr{S}_j}$$

Fig. 1

Huffman's procedure for finding an encoding of $\mathcal{S}_{j-1}$ is the following very easy rule.

Suppose the probabilities $q_1 \geq q_2 \geq \ldots \geq q_{t+1}$ of the words of $\mathcal{S}_{j-1}$ are such that the word formed from v_t and v_{t+1} in $\mathcal{S}_{j-1}$ is the word u_k of $\mathcal{S}_j$. Then the Huffman encoding of $\mathcal{S}_{j-1}$ would be as shown in the left-hand column of Fig. 1. Formally, it would be given by the rule

$$v_i \mapsto \sigma_i \quad (1 \leq i \leq k-1), \qquad v_i \mapsto \sigma_{i+1} \quad (k \leq i \leq t-1),$$

$$v_t \mapsto (\sigma_k, 0), \qquad v_{t+1} \mapsto (\sigma_k, 1).$$

Thus, working backwards, we start our encoding procedure by encoding the two words of $\mathcal{S}_{N-2}$ with the two code words 0 and 1; then $\mathcal{S}_{N-3}$ will have three code words, and so on, and we continue the above procedure until we arrive at the Huffman code for $\mathcal{S} = \mathcal{S}_0$.

We illustrate the method by a very small example.

Example Suppose $\mathcal{S}$ is a source with five source words and probabilities as shown. The development of a Huffman encoding can be seen by tracing the arrows forward and then the encoding backwards

W	P	C	W	P	C	W	P	C	W	P	C
w_1	0.5	1	v_1	0.5	1	u_1	0.5	1	x_1	0.5	0
w_2	0.2	01	v_2	0.2	01	u_2	0.3	00	x_2	0.5	1
w_3	0.15	001	v_3	0.15	000	u_3	0.2	01			
w_4	0.1	0000	v_4	0.15	001						
w_5	0.05	0001									

W: word,   P: probability,   C: code

The resulting encoding,

$$w_1 \rightarrow 1, \qquad w_2 \rightarrow 01, \qquad w_3 \rightarrow 001, \qquad w_4 \rightarrow 0000, \qquad w_5 \rightarrow 0001,$$

has average length 1.95 bits per source word. □

Note At least twice in the above example we were able to exercise choice, because two words had equal probabilities. When this occurs, we get distinct encodings.

PROOF THAT HUFFMAN'S ALGORITHM IS CORRECT

By Lemma 1, we know that the encoding of $\mathcal{S}_{N-2}$ by the two symbols 0 and 1 is optimum.

Hence, the proof will be complete if we can show that compactness is preserved as we move from the source $\mathscr{S}_j$ to $\mathscr{S}_{j-1}$ above.

Thus, let us assume that $\mathscr{S}_j$ is compactly encoded and that $l_1, \ldots, l_t$ are the lengths of the words $\sigma_1, \ldots, \sigma_t$ but that the Huffman encoding $\mathscr{C}_{j-1}$ of $\mathscr{S}_{j-1}$ is not compact. Hence there exists an instantaneous compact code $\mathscr{E}$ of $\mathscr{S}_{j-1}$ such that

$$\ell(\mathscr{E}) < \ell(\mathscr{C}_{j-1}).$$

By Lemma 3, we can rearrange the code words of $\mathscr{E}$ of maximum length to ensure that, if $\mathscr{E}$ has code words $v'_1, \ldots, v'_{t+1}$ with lengths $l'_1, \ldots, l'_{t+1}$ respectively, then $l'_1 \le l'_2 \le \ldots \le l'_{t+1}$ and

$$v'_t = (\sigma, 0), \qquad v'_{t+1} = (\sigma, 1),$$

where σ is some string from Σ^*.

Let the encoding $\mathscr{E}'$ of $\mathscr{S}_j$ consist of, in order, the words

$$v'_1, v'_2, \ldots, v'_{k-1}, \sigma, v'_k, \ldots, v'_{t-1};$$

then $\mathscr{E}'$ is an instantaneous encoding of $\mathscr{S}_j$ and has average length

$$\ell(\mathscr{E}') = p_1 l'_1 + \ldots + p_{k-1} l'_{k-1} + p_k |\sigma| + p_{k+1} l'_k + \ldots + p_t l'_{t-1}$$
$$= \ell(\mathscr{E}) - p_k.$$

But

$$\ell(\mathscr{C}_j) = \ell(\mathscr{C}_{j-1}) - p_k.$$

hence, if $\ell(\mathscr{E}) < \ell(\mathscr{C}_{j-1})$, then

$$\ell(\mathscr{E}') < \ell(\mathscr{C}_j)$$

which contradicts the assumption that $\mathscr{C}_j$ is compact. $\qquad\square$

HUFFMAN CODES OVER NON-BINARY ALPHABETS

Suppose instead of working with the alphabet $\{0, 1\}$ we have an alphabet Σ of r symbols.

Basically, the same construction works. As in the binary case, we start with $\mathscr{S}_0 = \mathscr{S}$ (the original source) and move through a sequence of sources $\mathscr{S}_0, \mathscr{S}_1, \ldots, \mathscr{S}_t$ until we arrive at a source $\mathscr{S}_t$ containing only r symbols. This has a compact encoding by just mapping the symbols 1–1 with Σ. There are only two points to watch:

(1) As we move from $\mathscr{S}_j$ to $\mathscr{S}_{j+1}$ collect not 2 but the r least probable symbols of $\mathscr{S}_j$ into one symbol of $\mathscr{S}_{j+1}$. Thus $\mathscr{S}_{j+1}$ has $r - 1$ fewer symbols than $\mathscr{S}_j$.

(2) We need the final source $\mathscr{S}_t$ to have exactly r symbols and hence, in order to achieve this, we need to start off with a source $\mathscr{S}$

of $r + t(r - 1)$ symbols. Since, in general, $\mathscr{S}$ is unlikely to have exactly this number of words, we artificially augment it by letting $\mathscr{S}_0 = \mathscr{S} \cup \mathscr{S}'$, where $\mathscr{S}'$ is a collection of dummy words of zero probability, with $|\mathscr{S}'| = r + t(r - 1) - |\mathscr{S}|$.

Exercises 2.5

1. What is the average word length of a compact code over $\{0, 1\}$ when there are five equiprobable source words?

2. Find the compact code over $\{0, 1\}$ for a source that emits words $w_1, \ldots, w_6$ with

$$P(w_1) = \tfrac{1}{3}, \qquad P(w_2) = \tfrac{1}{4}, \qquad P(w_3) = \tfrac{1}{6}, \qquad P(w_4) = P(w_5) = P(w_6) = \tfrac{1}{12},$$

and compare its average length with the upper and lower bounds given by the noiseless coding theorem.

3. Find a compact code over $\Sigma = \{0, 1, 2\}$ for the source of the previous example.

PROBLEMS 2

1. In a game on a chessboard one player (Algy) has to guess where his opponent has placed the Queen. Algy is allowed six questions which must be answered truthfully by a yes/no reply. Prove that there is a strategy by which Algy can always win this game, but that he cannot ensure winning is he is allowed only five questions.

2. If the game of the previous question is played on an $n \times n$ chessboard how many questions does Algy need in order to be certain of winning?

3. Find the average length of an optimum (i.e. compact) uniquely decipherable binary code for a zero-memory source that emits six words with probabilities

$$0.25, \quad 0.1, \quad 0.15, \quad 0.05, \quad 0.2, \quad 0.25.$$

By analysing the Huffman algorithm, show that, if a zero-memory source emits N words, and if $l_1, \ldots, l_N$ are the lengths of the codewords in an optimum encoding over the binary alphabet, then

$$l_1 + \ldots + l_N \leq \tfrac{1}{2} (N^2 + N - 2).$$

(Oxford 1986)

4. You are given a balance and nine apparently identical coins. You are told that one coin is different from the rest and asked to find which coin it is and whether it is heavier or lighter. Devise a strategy of three weighings to solve the problem.

In order to do the same problem for n coins in k weighings we need $k \log 3 \geq \log 2n$. Justify.

5. In Huffman's algorithm with the binary alphabet applied to N sourcewords, the lengths of the words in the final optimum encoding are

$l_1, \ldots, l_N$. Prove that

$$l_1 + \ldots + l_N \geq N \log_2 N.$$

6. Suppose that two persons whom we call Algy and the Oracle play a game with a random die. This die has n sides and takes values $1, \ldots, n$ with probabilities $p_1, \ldots, p_n$ respectively.

   The Oracle rolls the die behind a screen and Algy has to find out the value thrown as quickly as possible by asking questions of the Oracle. The Oracle always tells the truth but can only answer yes/no. Show that the average number of questions asked by Algy in any successful strategy must be at least the entropy $-\sum p_i \log p_i$.

7. Show that, in an optimum encoding of a source with N equiprobable source words in an alphabet of size D, there are exactly

$$\min\{(D^{r+1} - N - b)/(D - 1), N\}$$

   words of length r, where b is given by

$$N + b = D + k(D - 1), \qquad 0 \leq b < D - 1,$$

   and r is the greatest integer such that

$$D^r \leq N + b.$$

8. Show that the following method gives a uniquely decipherable code for a source $\mathcal{S}$.

   Suppose $\mathcal{S}$ has N sourcewords $s_1, s_2, \ldots, s_N$ and p_i is the probability that s_i is emitted, where the ordering is such that $p_i \geq p_{i+1}$. Let

$$a_1 = 0, \qquad a_2 = p_1, \qquad a_3 = p_1 + p_2, \ldots, \qquad a_N = p_1 + \ldots + p_{N-1}.$$

   Let m_i $(1 \leq i \leq N)$ be defined by:

$$m_j \text{ is the smallest integer satisfying } p_j \geq 2^{-m_j} \ (1 \leq j \leq N).]$$

   Then, if a_j^* is the binary expansion of a_j to m_j decimal places, the coding

$$s_j \mapsto a_j^* \quad (1 \leq j \leq N)$$

   is a uniquely decipherable code for $\mathcal{S}$. Show that it is not optimum, but that the average length $\hat{l}$ of the code satisfies

$$H(\mathcal{S}) \leq \hat{l} \leq H(\mathcal{S}) + 1.$$

   (Shannon, 1948)

9. If $l_1, \ldots, l_n$ are the lengths of words in a binary Huffman coding of sourcewords having probabilities $p_1, \ldots, p_n$, then the *redundancy* of the code is defined to be

$$r = \sum_{k=1}^{n} p_k l_k - H(p_1, \ldots, p_n).$$

Show that

$$r \leq p_{max} + \log[2(\log e)/e] = p_{max} + 0.086$$

where $p_{max} = \max_i p_i$.

(Gallager, 1978)

3

Communication through noisy channels

3.1 The discrete memoryless channel

In its broadest sense, a communication channel can be regarded as a black box that accepts strings of symbols from its input alphabet Σ_1 and emits strings of symbols from an output alphabet Σ_2.

Clearly, little of mathematical interest can be said about such a structure unless we put some further constraints on the way it works. Accordingly, we restrict attention to *discrete memoryless channels*, which are characterized by an input alphabet $\Sigma_1 = \{a_1, \ldots, a_m\}$, an output alphabet $\Sigma_2 = \{b_1, \ldots, b_n\}$, and a *channel matrix* $P = (p_{ij} : 1 \le i \le m, 1 \le j \le n)$. The channel's mode of operation is that, if any sequence $(u_1, u_2, \ldots, u_N)$ of symbols from Σ_1 is input to it, the output sequence is a string $(v_1, \ldots, v_N)$ of the same length with $v_k \in \Sigma_2$ and

$$P(v_k = b_j \mid u_k = a_i) = p_{ij} \quad (1 \le i \le m, 1 \le j \le n),$$

independently for each k.

Implicit in the above description is that the probabilities p_{ij} satisfy, for each i, the constraint

(1)
$$\sum_j p_{ij} = 1.$$

A matrix such as P with only non-negative entries and with row sums equal to 1 is called a *stochastic matrix*; those familiar with random processes will recognize P as the transition matrix of a Markov chain.

It is often useful to represent a channel by a diagram, as shown in Example 1.

Example 1 The *binary erasure channel* has $\Sigma_1 = \{0, 1\}$, output alphabet $\Sigma_2 = \{0, 1, *\}$, and channel matrix P:

$$
P = \begin{array}{c} \\ 0 \\ 1 \end{array}
\begin{array}{ccc}
0 & 1 & * \\
\left[\begin{array}{ccc} 1 - \varepsilon & 0 & \varepsilon \\ 0 & 1 - \varepsilon & \varepsilon \end{array}\right]
\end{array}
$$

a diagrammatic form of the channel being given on the right. This corresponds to a situation in which any symbol has probability ε of being fudged, and hence being output as *. But neither 0 nor 1 has any chance of being transposed into the other. □

Example 2

The most commonly used channel in this model of communication is the *binary symmetric channel*. This is a discrete memoryless channel with input and output alphabets $\Sigma = \{0, 1\}$ and with channel matrix P given by

$$P = \begin{bmatrix} 1-p & , & p \\ p & , & 1-p \end{bmatrix}$$

In other words, there is a common probability p of any symbol being transmitted incorrectly, independently for each symbol transmitted. Very often we will write $q = 1 - p$ without further mention. □

EXTENSIONS OF A DISCRETE MEMORYLESS CHANNEL

Consider a discrete memoryless channel with input alphabet Σ_1, output alphabet Σ_2, and channel matrix P. The *r-th extension* of this channel is the discrete memoryless channel with input alphabet $\Sigma_1^{(r)}$, output alphabet $\Sigma_2^{(r)}$, and channel matrix $P^{(r)}$, where $P^{(r)}$ is defined as follows.

The (i, j)-entry of $P^{(r)}$, corresponding to an input

$$\sigma_i = \alpha_1 \alpha_2 \ldots \alpha_r,$$

with $\alpha_k \in \Sigma_1$, and output

$$\tau_j = \beta_1 \beta_2 \ldots \beta_r,$$

with $\beta_k \in \Sigma_2$, is

$$(P^{(r)})_{ij} = p(\beta_1 \mid \alpha_1) p(\beta_2 \mid \alpha_2) \ldots p(\beta_r \mid \alpha_r)$$

where $p(\beta_k \mid \alpha_k)$ is the probability that, in the channel with matrix P, the symbol β_k is received when α_k is transmitted.

Example 3

The second extension of the binary symmetric channel with matrix

$$P = \begin{bmatrix} q & p \\ p & q \end{bmatrix}$$

has input and output alphabets $\{00, 01, 10, 11\}$ and channel matrix

$$P^{(2)} = \begin{bmatrix} q^2 & qp & pq & p^2 \\ qp & q^2 & p^2 & pq \\ pq & p^2 & q^2 & qp \\ p^2 & pq & qp & q^2 \end{bmatrix} = \begin{bmatrix} qP & pP \\ pP & qP \end{bmatrix}.$$

□

An alternative and instructive way to think of an rth extension of a channel C is to regard it as r copies of C operating independently and in parallel as shown in Fig. 2.

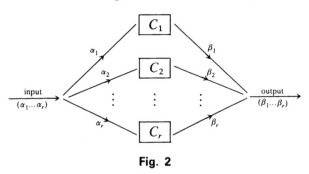

Fig. 2

Exercise 3.1   1. A message consisting of N binary digits is transmitted through a binary symmetric channel having error probability p. Show that the expected number of errors is Np.

3.2 Connecting the source to the channel

Consider the following situation: we have a memoryless source $\mathscr{S}$ which emits symbols (or source words) $s_1, \ldots, s_N$ with probabilities $p_1, \ldots, p_N$. This source is connected to a binary symmetric channel with error probability p as shown:

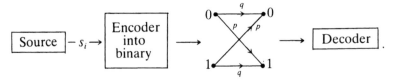

We assume that the encoding into binary is noiseless, and is known to the decoder.

Suppose for simplicity that $N = 8$; we may think of the symbols as the names of racehorses, different currencies, letters of the alphabet, or whatever. An efficient (i.e. compact) encoding in the sense of the previous chapter would be

(1)

$$
\begin{array}{cccccccc}
s_1 & s_2 & s_3 & s_4 & s_5 & s_6 & s_7 & s_8 \\
\downarrow & \downarrow & \downarrow & \downarrow & \downarrow & \downarrow & \downarrow & \downarrow \\
000 & 100 & 010 & 001 & 110 & 101 & 011 & 111.
\end{array}
$$

A *message* is a string of source words s_i that are successively encoded, transmitted, and then decoded. Hence, with the particular encoding scheme (1), the probability that any particular word is correctly transmitted is q^3. Thus, the probability that a message of n words is correctly transmitted is q^{3n}.

Can we do better? The answer is obviously 'yes', or we would not be asking the question. The interesting questions are (a) how much better and (b) at what cost?

Example   Consider the above example with eight equiprobable source words and suppose we 'double up' the encoding and use

$$
\begin{array}{cccccccc}
s_1 & s_2 & s_3 & s_4 & s_5 & s_6 & s_7 & s_8 \\
\downarrow & \downarrow & \downarrow & \downarrow & \downarrow & \downarrow & \downarrow & \downarrow \\
000\,000 & 100\,100 & 010\,010 & 001\,001 & 110\,110 & 101\,101 & 011\,011 & 111\,111.
\end{array}
$$

So long as our decoder adopts the rule that it only decodes when the first three symbols and second three symbols agree, and otherwise 'yells for help', the probability that an error occurs *and* remains undetected is drastically reduced. Of course, we are paying a considerable price: we are reducing the rate of transmission by a factor of 2. Moreover, this is purely a detection system which will be of no use if the decoder can't contact the encoder and ask for the word detected to be in error to be repeated.   □

The rest of this chapter is concerned with achieving reliable transmission through a noisy channel without sacrificing too much and in the situation where the receiver has no feedback to the sender.

Exercise   1. A simple way of *detecting* up to one error is to use the device of adding a
3.2   *parity check* to ensure that the sum of the digits in a transmitted word is
              even. Thus a parity-check encoding of (1) above would be

$$
s_1 \rightsquigarrow 0000, \quad s_2 \rightsquigarrow 1001, \quad \ldots, \quad s_8 \rightsquigarrow 1111.
$$

Show that, if such a parity-check code is transmitted through a binary symmetric channel, the probability that an error is not noticed is given by $6p^2(1-p)^2 + p^4$, where p is the probability of error in the channel.

3.3 Codes and decoding rules

Given a memoryless channel with input alphabet Σ_1 and output alphabet Σ_2, a *code* of *length* n is any collection $\mathscr{C}$ of distinct n-sequences of symbols from Σ_1. The elements of $\mathscr{C}$ are called the *codewords*. Given a code $\mathscr{C}$ of length n with codewords $c_1, \ldots, c_N$, a *decoding rule* is any partition of the set of possible received sequences into disjoint sets $R_1, \ldots, R_N$ with the obvious interpretation that, if the received sequence y is a member of R_j, it is decoded as the code word c_j.

The choice of a decoding rule is crucial to the success of any communication system. As an extreme example, it is easy (by being wilfully stupid) to construct a decoding rule which completely destroys the reliability of a perfectly reliable (that is, noiseless) channel.

Example 1
Suppose that we have a source with just two source words s_1 and s_2, which we encode for transmission across a binary symmetric channel by

$$s_1 \mapsto 000 = c_1, \qquad s_2 \mapsto 111 = c_2.$$

There are eight possible received messages. A possible decoding rule might be to decode a message as s_1 only if it contained more zeros than ones. A less sensible rule might be to decode a message as s_1 only when the received message was 000. A priori, each of these rules has equal standing, as does the rule: decode every received message as s_1!   □

Which decoding rule should we use? The obvious answer is to try to minimize the chance of error. This suggests decoding any received vector y into a code word c_j such that

(1) $$P(c_j \text{ sent} \mid y \text{ received}) \geq P(c_i \text{ sent} \mid y \text{ received}) \quad \text{for all } i.$$

This is called the *ideal-observer* or *minimum-error* rule. However, rewriting these conditional probabilities shows that this rule cannot be used without knowledge of the probabilities with which the c_j are used.

In practice, this is a serious disadvantage. This, together with the fact that it is not that easy to use when there are a large number of codewords, justifies the use of the following rule called the *maximum-likelihood* (ML) decoding rule. This works by decoding a received y into a c_j that maximizes

(2) $$P(y \text{ received} \mid c_j \text{ sent}).$$

To those familiar with maximum-likelihood estimators in statistics, the analogy is obvious.

In the absence of any information about the probabilities of different code words, we note also:

(3)   If the code words are equally probable, then the maximum-likelihood decoding rule agrees with the ideal-observer rule.

The proof is very easy and is left to the reader.

HAMMING DISTANCE

For the main part of this course, we shall be dealing with the binary symmetric channel. For this channel the maximum likelihood decoding rule has a particularly easy implementation.

We will let V_n denote the set of all n-sequences of 0s and 1s and, when necessary, will regard V_n as an n-dimensional vector space over the field of integers modulo 2. If x and y are vectors in V_n we define the *Hamming distance* $d(x, y)$ between x and y to be the number of places in which x and y differ.

For the binary symmetric channel, a natural decoding rule is the *minimum-distance* rule, namely: decode any received vector y into a codeword c that is a minimum Hamming distance from y; if there is more than one such codeword choose one arbitrarily.

A very easy result is the following.

Theorem 1   *For the binary symmetric channel with probability of error $p \leq \frac{1}{2}$, the minimum-distance decoding rule is equivalent to the maximum-likelihood decoding rule.*

Proof   For any vectors x and y of V_n with $d(x, y) = d$,

$$P(y \text{ received} \mid x \text{ sent}) = p^d q^{n-d}.$$

When $p < \frac{1}{2}$, this is a maximum when d is a minimum.   □

Exercises 3.3   1. A code consists of 4 codewords $c_1 = 1000$, $c_2 = 0110$, $c_3 = 0001$, and $c_4 = 1111$. The probabilities that these codewords occur are given by

$$P(c_1) = P(c_2) = \tfrac{1}{3}, \qquad P(c_3) = P(c_4) = \tfrac{1}{6}.$$

If you are using the binary symmetric channel with probability of error given by $\frac{1}{10}$, and you receive the vector 1001, how would you decode it (a) using the ideal-observer rule and (b) using the maximum-likelihood rule?

2. Prove statement (3) that, when the codewords are equiprobable, the maximum-likelihood rule agrees with the ideal-observer rule.

3.4 The capacity of a channel

As its name suggests, the capacity of a communication channel is a measure of its ability to transmit information. The formal definition is motivated as follows.

Suppose that we have a discrete memoryless channel with input alphabet $\Sigma_1 = \{a_1, \ldots, a_m\}$ and output alphabet $\Sigma_2 = \{b_1, \ldots, b_n\}$, together with channel matrix

$$P = [p_{ij}] = P(b_j \text{ received} \mid a_i \text{ sent}).$$

If we attach to this channel a memoryless source $\mathscr{S}$ which emits the symbols $a_1, \ldots, a_m$ with probabilities $p_1, \ldots, p_m$, then the output of the channel could be regarded as a memoryless source $\mathscr{J}$ which emits the symbols $b_1, \ldots, b_n$ with probabilities $q_1, \ldots, q_n$, where

$$q_j = \sum_{i=1}^{m} P(b_j \text{ received} \mid a_i \text{ sent}) P(a_i \text{ sent})$$

$$= \sum_{i=1}^{m} p_i p_{ij}.$$

The information about $\mathscr{S}$ given by $\mathscr{J}$ is, as defined in Chapter 1, $I(\mathscr{S} \mid \mathscr{J})$ where

$$I(\mathscr{S} \mid \mathscr{J}) = H(\mathscr{S}) - H(\mathscr{S} \mid \mathscr{J}) = H(\mathscr{S}) + H(\mathscr{J}) - H(\mathscr{S}, \mathscr{J})$$

is a function only of the source distribution $(p_1, \ldots, p_m)$ and the channel matrix P. It is natural therefore to define the *capacity* C of the channel by

(1)
$$C = \sup I(\mathscr{S} \mid \mathscr{J}),$$

where the supremum is taken over all memoryless sources $\mathscr{S}$ or, more precisely, over all possible input distributions $(p_1, \ldots, p_m)$.

First, note that C is well defined in the sense that all we are seeking is the supremum of $f(\boldsymbol{p})$, where f is a continuous function on a closed and bounded subset of $\mathbb{R}^m$ and, by a fundamental theorem of analysis, any continuous function on such a set attains its supremum on the set. Thus (1) can be rewritten as

(1)
$$C = \max I(\mathscr{S} \mid \mathscr{J}).$$

Secondly, note that C is a quantity determined entirely by the channel matrix P. It is roughly analogous to the conductance of a resistor in electrical network theory. Its units are clearly the units of information or entropy, namely *'bits per second'* or *'bits per symbol'*, depending on the context.

To give an example of how capacities are found, we prove the following theorem.

Theorem 1   *The capacity of the binary symmetric channel in which there is probability p of error is given by*

(2)
$$C(p) = 1 + p \log p + q \log q,$$

where $q = 1 - p$.

Proof   To ease the notation, suppose that the source emits 0 with probability α and 1 with probability $\beta = 1 - \alpha$. Thus the output $\mathcal{J}$ has distribution

0 with probability $\alpha q + \beta p$,   1 with probability $\beta q + \alpha p$.

Then $H(\mathcal{S}, \mathcal{J})$ is just the entropy of the distribution $(\alpha q, \alpha p, \beta q, \beta p)$. Thus, after a little manipulation, we obtain

$$I(\mathcal{S} \mid \mathcal{J}) = p \log p + q \log q - (\alpha q + \beta p)\log(\alpha q + \beta p)$$
$$- (\alpha p + \beta q)\log(\alpha p + \beta q).$$

Differentiating with respect to α (remembering that $\beta = 1 - \alpha$) shows that this has a maximum when $\alpha = \frac{1}{2}$ and gives (1).   □

Notice that capacity has the expected properties that $C(p)$ is a monotone function of p $(0 \le p \le \frac{1}{2})$ and

$$C(0) = 1, \qquad C(\tfrac{1}{2}) = 0,$$

which agrees with the intuitive notions that, when $p = \frac{1}{2}$, the channel becomes a perfect scrambler but, when $p = 0$, we have perfect transmission.

Working out the capacity of general channels is nontrivial. Unless the channel has some special property or is related to a channel whose capacity is already known, the only way to evaluate the capacity is to solve a problem in constrained optimization, typically by using the method of Lagrange multipliers.

As an example of the first of these techniques, we prove the following result.

Theorem 2   *If a memoryless channel has capacity C, then its r-th extension has capacity rC.*

Proof   Denote by $C^{(r)}$ the capacity of the extension so that

(3)
$$C^{(r)} = \sup_X H(X) - H(X \mid Y),$$

where $X = (X_1, \ldots, X_r)$ and $Y = (Y_1, \ldots, Y_r)$ are an input/output pair. Write

(4)
$$H(X) - H(X \mid Y) = H(Y) - H(Y \mid X).$$

But, in the obvious notation, we have

$$H(Y \mid X) = \sum_x p(x) H(Y \mid X = x).$$

Since the channel is memoryless,

$$H(Y \mid X = x) = \sum_i H(Y_i \mid X = x) = \sum_i H(Y_i \mid X_i = x_i).$$

Thus

$$H(Y \mid X) = \sum_x p(x) \sum_i H(Y_i \mid X_i = x_i)$$
$$= \sum_i \sum_u H(Y_i \mid X_i = u) P(X_i = u).$$

Hence,

(5)
$$H(Y \mid X) = \sum_{i=1}^{r} H(Y_i \mid X_i).$$

Now, from (1.2.4) we get that

$$H(Y) \le H(Y_1) + \ldots + H(Y_r),$$

and so, from (3), (4), and (5), we get $C^{(r)} \le rC$.

We note that equality will hold only if the $Y_1, \ldots, Y_r$ are independent. This can be achieved by taking $X_1, \ldots, X_r$ independent and choosing, the distribution that achieved the capacity C of the underlying channel. ☐

Exercises 3.4
1. Calculate the capacity of the binary erasure channel of error probability ε.
2. By considering the memoryless channel with matrix

$$\begin{bmatrix} 1 & 0 \\ 0 & 1 \\ 0 & 1 \end{bmatrix},$$

show that capacity may be achieved by more than one input distribution. Show also that its second-order extension can achieve capacity with an input distribution that is not the product of input distributions to the original channel.

(Feinstein, 1958)

3.5   The noisy coding theorem

We saw earlier that we can achieve arbitrarily high reliability by just repeating each source symbol sufficiently often. Obviously this is very time-consuming, and the main thrust of this chapter is to prove a marvellous theorem of C. Shannon (1948) which shows that, provided one keeps the transmission rate below the channel capacity, we can achieve arbitrarily high reliability. We will concentrate attention on the binary symmetric channel. The ideas extend to more complicated channels, but it is more important to understand fully the underlying principles than to get bogged down in mathematical details.

Given any code $\mathscr{C}$ and any decoding scheme for $\mathscr{C}$, the *error probability* $e(\mathscr{C})$ is usually defined as the average probability of error on the assumption that all codewords are equally likely to have been sent. In other words, if there are M codewords $c_1, \ldots, c_M$ in $\mathscr{C}$, then

$$e(\mathscr{C}) = \frac{1}{M} \sum_{i=1}^{M} P(\text{error} \mid c_i \text{ transmitted}).$$

In the case of binary codes, we assume, unless otherwise specified, that the maximum-likelihood (=minimum-distance) decoding rule is being used, and hence often refer to the error probability of a code without any specific mention of the decoding rule.

Obviously, the object of the exercise is to find codes with small average error probability. However, an even more stringent requirement is to demand that the *maximum error probability* is small. As expected, this is defined by

$$\hat{e}(\mathscr{C}) = \max_{i} P(\text{error} \mid c_i \text{ transmitted}),$$

and clearly

$$\hat{e}(\mathscr{C}) \geq e(\mathscr{C}).$$

Suppose therefore that we have a binary symmetric channel which has error probability p and hence capacity C given by

$$C = C(p) = 1 + p \log p + (1-p)\log(1-p).$$

We shall prove the following version of Shannon's noisy coding theorem.

Theorem 1

Given a binary symmetric channel of capacity C and any R, with $0 < R < C$, then, if $(M_n : 1 \leq n < \infty)$ is any sequence of integers satisfying

$$1 \leq M_n \leq 2^{Rn} \quad (1 \leq n < \infty),$$

and $\varepsilon > 0$ any given positive quantity, there exists a sequence of codes $(\mathscr{C}_n : 1 \leq n < \infty)$ and an integer $N_0(\varepsilon)$ with $\mathscr{C}_n$ having M_n codewords of length n and with maximum error probability

$$\hat{e}(\mathscr{C}_n) \leq \varepsilon$$

for all $n \geq N_0(\varepsilon)$.

How does the theorem work? Well, suppose that the error probability is such that the channel has capacity $C(p) = 0.8$. Then, if our message is a string of 0's and 1's, we know that, for sufficiently large n, if we take $R = 0.75$, there exists a set of $2^{0.75n}$ codewords of length n that have error probability less than any prescribed threshold. Hence, in order to encode the message stream from the source, the procedure would be:

(a) Break the message stream up into blocks of length m, where m is such that $3\lceil \frac{1}{4}n \rceil = m \geq \frac{3}{4}N_0(\varepsilon)$.

(b) Encode these m-blocks into the code $\mathscr{C}_n$ using a codeword of length $\frac{4}{3}m$ for each m-block.

(c) Transmit the new encoded stream through the channel.

What have we achieved?: A marked reduction in the error probability. At what cost?: Complexity of encoding and slower rate of transmission: also we don't yet know the code. The thrust of Shannon's theorem is to tell us that such codes exist.

The proof of Shannon's theorem that we shall give below depends on two inequalities. The first is well known; its proof can be found in any elementary text on probability.

CHEBYSHEV'S INEQUALITY

If X is any random variable with finite variance $\text{var}(X)$, then, for any $a > 0$,

(1) $$P(|X - E(X)| \geq a) \leq \text{var}(X)/a^2.$$

The second inequality is less well known. It too has a probabilistic interpretation; we state it as follows.

THE TAIL INEQUALITY

For any λ, with $0 \le \lambda \le \frac{1}{2}$,

(2)
$$\sum_{k=0}^{\lfloor \lambda n \rfloor} \binom{n}{k} \le 2^{nh(\lambda)},$$

where $h(\lambda) = -[\lambda \log \lambda + (1 - \lambda)\log(1 - \lambda)]$.

Proof For ease of presentation, and with no loss of generality, we may assume λ is such that λn is an integer. Then we can write

$$1 = [\lambda + (1 - \lambda)]^n \ge \sum_{k=0}^{\lambda n} \binom{n}{k} \lambda^k (1 - \lambda)^{n-k}$$

$$\ge (1 - \lambda)^n \sum_{k=0}^{\lambda n} \binom{n}{k} \left(\frac{\lambda}{1 - \lambda}\right)^{\lambda n}$$

$$= \lambda^{\lambda n}(1 - \lambda)^{n(1-\lambda)} \sum_{k=0}^{\lambda n} \binom{n}{k}.$$

Hence,

$$\sum_{k=0}^{\lambda n} \binom{n}{k} \le \lambda^{-\lambda n}(1 - \lambda)^{-n(1-\lambda)}.$$

Taking logarithms to base 2 gives (2). $\qquad\square$

PROOF OF THE NOISY CODING THEOREM

We first describe the broad outline of the proof. Fix an integer n and, for the moment, work with binary codes in V_n. Suppose that we are trying to find a code with M codewords $c_i \in V_n$. We will choose these codewords c_i by the somewhat crazy method of just selecting a vector at random from V_n, independently for each i $(1 \le i \le M)$. This is called *random coding*.

We will decode by the following method: fix $r > 0$ and let $S_r(y)$ denote the *r-sphere* about y, that is

$$S_r(y) = \{z : z \in V_n, d(y, z) \le r\}.$$

Then, if y is the received vector, we decode y as the codeword c_j if c_j is the unique codeword in $S_r(y)$; otherwise decode y as some arbitrary codeword, perhaps c_1.

We now begin the detailed proof. Let Y denote the vector received when a codeword c is transmitted and let E be the event that an error is made, that is, that Y is decoded as a codeword other than c. Now, an error can only occur if either

(a) $d(c, Y) > r$

or

(b) $d(c, Y) \leq r$ and $d(c', Y) \leq r$ for some other codeword c'.

Denote the events described by (a) and (b) by A and B respectively, so that $E = A \cup B$, and hence

$$P(E) = P(A \cup B) \leq P(A) + P(B).$$

Consider B; this will occur if both
 (i) not more than r errors are made in transmission
and
(ii) one of the codewords other than c is within distance r of the received vector Y.

If these two events are denoted by B_1 and B_2 respectively, then, since $B = B_1 \cap B_2$, we have

(3) $$P(B) \leq P(B_2).$$

Now consider B_2: since the codewords are chosen randomly, the probability that c_i (say) is within a distance r of Y is $N_r(n)/2^n$, where

(4) $$N_r(n) = \sum_{k=0}^{r} \binom{n}{k}$$

counts the number of vectors in V_n that belong to $S_r(y)$. Hence the probability that at least one of the other $M - 1$ codewords (not equal to c) is within a distance r of the received word Y satisfies

(5) $$P(B_2) \leq \frac{(M-1)}{2^n} \sum_{k=0}^{r} \binom{n}{k}.$$

Hence if, for any $\varepsilon > 0$, we take

$$r = \lfloor np + n\varepsilon \rfloor$$

to be the maximum integer not greater than $np + n\varepsilon$, we get from (3), (4), and the tail inequality (2) that

(6) $$P(B) \leq \frac{M}{2^n} 2^{nh(p+\varepsilon)}.$$

Turning now to the second type of error given by A, notice that, if

U is the (random) number of symbols in error when transmitting the codeword c, then

$$P(A) = P(U > r)$$

and U is a binomial random variable with parameters n and p. Hence

$$P(A) = P(U > np + n\varepsilon) \le P(|U - np| > n\varepsilon)$$
$$\le \text{var}(U)/n^2\varepsilon^2,$$

by Chebyshev's inequality.

Because U is a binomial random variable,

$$\text{var}(U) = npq$$

and hence the total probability of error satisfies

$$P(E) \le M2^{-n[1-h(p+\varepsilon)]} + pq/n\varepsilon^2,$$

for sufficiently large n. Since $C(p + \varepsilon) = 1 - h(p + \varepsilon)$, this reduces to

$$P(E) \le \frac{pq}{n\varepsilon^2} + M2^{-nC(p+\varepsilon)}.$$

Since $\varepsilon > 0$, this error probability can be made arbitrarily small for sufficiently large n, provided that M, regarded as a function of n, grows at a rate no faster than $2^{nC(p)}$.

Hence we have proved the noisy coding theorem as stated, except that we have bounded the *average* error probability rather than the maximum error probability. To complete the proof we need to show the existence of codes $\mathscr{C}_n$ with M_n codewords, where $M_n \le 2^{Rn}$, and with maximum error probability $< \varepsilon$. Accordingly take $\varepsilon' = \frac{1}{2}\varepsilon$ and $M'_n = 2M_n$, and note that, since $M_n \le 2^{Rn}$ and $R < C$, there must exist R', with $R < R' < C$, and N'_0 such that, for $n \ge N'_0$,

$$M'_n \le 2^{nR'}$$

and there exists a sequence of codes $\mathscr{C}'_n$ such that $\mathscr{C}'_n$ has M'_n codewords and *average* error probability $< \varepsilon'$ for $n \ge N'_0$.

If $x_1, \ldots, x_{M'_n}$ are the codewords of $\mathscr{C}'_n$, this means that

$$\sum_{i=1}^{M'_n} P(E \mid x_i) \le \varepsilon' M'_n.$$

Hence at least half of these codewords x_i must satisfy

(7)
$$P(E \mid x_i) \le 2\varepsilon' = \varepsilon.$$

Let $\mathscr{C}_n$ be any M_n of these codewords satisfying (7); then we have our required code with maximum error probability $\le \varepsilon$. $\square$

Shannon's theorem extends to the general memoryless channel with arbitrary input and output alphabets. The main idea of the proof is the same, namely

(a) code messages randomly

(b) decode by what is essentially the maximum-likelihood decision procedure.

Technical difficulties are caused mainly by the very general form of the capacity of the channel when it is not binary symmetric. The interested reader can find a complete proof (in fact two) for this general situation in Ash (1965) or Gallager (1968).

We should also mention, at this stage, the importance of improving the error-probability bounds. In our proof above, we were only interested in convincing the reader that the probability of error can be made arbitrarily small. A fairly extensive and highly technical literature exists detailing orders of magnitude for the rate at which these probabilities approach zero.

For example the following stronger result is due to Shannon (1957).

Theorem 2 *If a discrete memoryless channel has capacity $C > 0$ and R is any positive quantity with $R < C$, there exists a sequence of codes $(\mathscr{C}_n : 1 \leq n < \infty)$ such that:*

(a) *$\mathscr{C}_n$ has $\lfloor 2^{Rn} \rfloor$ codewords of length n;*

(b) *$\hat{e}(\mathscr{C}_n)$ the maximum error probability of $\mathscr{C}_{n'}$, satisfies*

$$\hat{e}(\mathscr{C}_n) \leq A e^{-Bn},$$

where A and B depend only on the channel and on R.

In other words, not only do there exist good codes but there are codes whose error probabilities decrease exponentially.

The proof of such a result is fairly technical, it depends essentially on using exponential bounding theorems for the law of large numbers and is beyond the scope of this course.

Exercises 3.5

1. A binary symmetric channel with error probability 0.05 can transmit 800 binary digits per second. How many bits can it transmit accurately per second?

2. A binary symmetric channel with a physical capability of transmitting 800 digits per second can transmit 500 digits per second with arbitrarily small error probability. What does this tell you about the error probability of the channel?

3.6   Capacity is the bound to accurate communication

Suppose we have a discrete memoryless channel of capacity C bits. Suppose also that the channel has a mechanical speed of 1 bit per second. We shall now prove a converse to Shannon's theorem by showing that it is impossible to transmit information accurately at a speed of more than C bits per second. More precisely, we shall prove the following fundamental result.

Theorem 1   *For a memoryless channel of capacity C and any $R > C$, there cannot exist a sequence of codes $(\mathscr{C}_n : 1 \le n < \infty)$, with the property that $\mathscr{C}_n$ has 2^{nR} codewords of length n and error probability $e(\mathscr{C}_n)$, that tends to zero as $n \to \infty$.*

In fact, a much stronger result was proved by Wolfowitz (1961), namely that, under the same conditions, the maximum error probabilities converge to 1 as $n \to \infty$. However, the weaker version given here suffices to show that Shannon's noisy coding theorem is best possible. In order to prove the theorem we need the following lemma.

FANO'S INEQUALITY

Let $\mathscr{C}$ be any code with M codewords $\{c_1, \ldots, c_M\}$ for a given discrete memoryless channel. Let X be a random vector taking values in the set of codewords. Let Y denote the (random) vector output when X is transmitted through the channel and decoded. Then, if p_E is the probability of an error (namely $p_E = P(X \ne Y)$), we have

(1)
$$H(X \mid Y) \le H(p_E, q_E) + p_E \log(M - 1),$$

where $q_E = 1 - p_E$.

Before proving Fano's inequality, we note that it has a very natural interpretation. The right-hand side is the sum of two terms. The first term $H(p_E, 1 - p_E)$ is the information needed to decide whether or not there is an error. The second term is the information needed to resolve the error, assuming the worst possible case. Unfortunately, this heuristic argument seems to need the following nontrivial proof.

Proof   First note that we make no use of the actual distribution of X in the proof; notice also that the right-hand side of the inequality (1) is independent of it. It is included in the statement purely for reasons of clarity. We will let x denote a typical codeword and y the decoding of the corresponding output.

We first prove a preliminary lemma.

Lemma *For any three random vectors U, V, W*

$$H(U \mid V) \le H(U \mid V, W) + H(W)$$

Proof Using the fundamental identity, Theorem 1.3.2, we can write

$$H(U \mid V) = H(U, V) - H(V)$$
$$= H(U, V, W) - H(W \mid U, V) - H(V)$$
$$\le H(U, W \mid V)$$

since entropy is non-negative. But

$$H(U, W \mid V) = H(U, V, W) - H(V, W) + H(V, W) - H(V)$$
$$= H(U \mid V, W) + H(W \mid V)$$
$$\le H(U \mid V, W) + H(W)$$

as required. □

Turning now to a proof of Fano's inequality introduce a new random variable Z defined by

$$Z = \begin{cases} 0 & \text{if } X = Y \\ 1 & \text{if } X \ne Y. \end{cases}$$

Thus

(2) $$H(Z) = H(p_E, q_E).$$

Consider now the pair (Y, Z). Clearly

$$H(X \mid (Y, Z) = (y, 0)) = 0.$$

Also, if $(Y, Z) = (y, 1)$, X is distributed over the $M - 1$ codewords which are not equal to y. Hence

$$H(X \mid (Y, Z) = (y, 1)) \le \log(M - 1).$$

Thus

(3) $$H(X \mid Y, Z) = \sum_y H(X \mid (Y, Z) = (y, 1))P((Y, Z) = (y, 1)$$
$$\le \log(M - 1) \sum_y P((Y, Z) = (y, 1))$$
$$= p_E \log(M - 1).$$

Taking $U = X$, $V = Y$, $W = Z$ in Lemma 1 and using (2) and (3) gives Fano's inequality.   ☐

Now we can prove our main result.

Proof of Theorem 1

Suppose such a sequence of codes exists. Then take X in Fano's Lemma to be equidistributed in the code $\mathscr{C}_n$ so that, if we take $R = C + \varepsilon$, with $\varepsilon > 0$, we have

$$H(X) = n(C + \varepsilon).$$

Since the capacity of the channel is C, then, for codewords of length n, we know it has capacity nC (the nth extension of a memoryless channel) and thus, denoting by Y the output corresponding to X, we get

$$nC \geq H(X) - H(X \mid Y),$$

which gives

$$H(X \mid Y) \geq n(C + \varepsilon) - nC = n\varepsilon.$$

Now using Fano's inequality we get, since there are $2^{n(C+\varepsilon)}$ codewords,

$$n\varepsilon \leq H(X \mid Y) \leq H(p_E, q_E) + p_E n(C + \varepsilon),$$

which gives

$$p_E \geq \frac{n\varepsilon - H(p_E, q_E)}{n(C + \varepsilon)},$$

and, as $n \to \infty$, this does not tend to zero. Thus the sequence of codes $\mathscr{C}_n$ cannot exist.   ☐

PROBLEMS 3

1. In the binary symmetric channel with error probability $\varepsilon > 0$, the code used consists of the two codewords 000 and 111. Show that, if the maximum-likelihood decoding rule is used, then the probability of error is $3\varepsilon^2 - 2\varepsilon^3$.
2. An *error burst* of length k consists of a sequence of k symbols, each of which is transmitted incorrectly. Find the expected number of error bursts of length k, when a message of length N is transmitted down a binary symmetric channel of error probability p.
3. A code for a binary symmetric channel, which has error probability $\varepsilon > 0$, consists of all 5-vectors over $\{0, 1\}$ that have exactly two ones.

What is the probability that the codeword 11000 is decoded as 10001 under minimum-distance decoding?

4. A number N of binary symmetric channels, each with error probability p, are connected in series. Show that the capacity C_N of the cascaded combination is given by

$$C_N = 1 + p_N \log p_N + q_N \log q_N,$$

where $p_N = \frac{1}{2}[1 - (p - q)^N]$ and $q_N = 1 - p_N$.

5. Consider two discrete memoryless channels, of capacities C_1 and C_2 and each having input alphabet Σ_1 and output alphabet Σ_2. The *product* of these channels is a channel whose input and output alphabets are $\Sigma_1^{(2)}$ and $\Sigma_2^{(2)}$ respectively, with channel probabilities given by

$$p(y_1 y_2 \mid x_1 x_2) = p_1(y_1 \mid x_1) p_2(y_2 \mid x_2),$$

where $p_i(y \mid x)$ is the probability that y is received when x is transmitted through the ith channel. Show that the capacity C of the product channel is given by

$$C = C_1 + C_2.$$

(Shannon, 1957)

6. A memoryless source $\mathscr{S}$ is connected to a channel C_1, and the resulting output $\mathscr{S}_1$ is input to a channel C_2 giving an output $\mathscr{S}_2$ as shown in the diagram below.

Show that, in the obvious notation,

$$I(\mathscr{S} \mid \mathscr{S}_2) \le I(\mathscr{S} \mid \mathscr{S}_1) \quad \text{and} \quad I(\mathscr{S} \mid \mathscr{S}_2) \le I(\mathscr{S}_1 \mid \mathscr{S}_2).$$

(This is one version of what is known as the data processing theorem, it says that communication cannot be improved by passing material through more processors—at the intuitive level it is obvious.)

7. Give an example to show that equality can hold in Fano's inequality.

8. If a channel matrix, with output alphabet of size n, is such that the set of entries in any row is the set $\{p_1, \ldots, p_n\}$, and the set of entries in each column is the same, show that its capacity C is given by

$$C = \log n + \sum_{i=1}^{n} p_i \log p_i.$$

Hence show that the capacity of the channel that has matrix

$$\begin{bmatrix} \frac{1}{3} & \frac{1}{3} & \frac{1}{6} & \frac{1}{6} \\ \frac{1}{6} & \frac{1}{6} & \frac{1}{3} & \frac{1}{3} \end{bmatrix}$$

is given by $C = \log 2^{\frac{5}{3}} - \log 3$.

(Shannon, 1948)

9. Two binary vectors of length n are chosen at random. What is the probability that their Hamming distance is at least k?

10. (For the microcomputer.) Let V_n denote the set of binary n-tuples. If $w_1, \ldots, w_k$ denote k codewords chosen at random from V_n, then denote by D_k the minimum distance between these codewords. Thus D_k is a random variable whose distribution depends on k and n. Finding even the expected value of D_k seems to be a very messy calculation, and is nontrivial even for $k = 3$. Estimate $E(D_k)$ as k and n vary, by using Monte Carlo methods. (This problem epitomizes the principle behind Shannon's randomized-coding proof of his main theorem.)

11. Consider the binary symmetric erasure channel which has channel matrix

$$\begin{bmatrix} 1 - \alpha - \beta & \alpha & \beta \\ \alpha & 1 - \alpha - \beta & \beta \end{bmatrix}.$$

Show that its capacity is

$$C = (1 - \beta)[1 - \log(1 - \beta)] + (1 - \alpha - \beta)\log(1 - \alpha - \beta) + \alpha \log \alpha.$$

12. Prove that, if $\frac{1}{2} < \lambda < 1$, then

$$\sum_{k = \lceil \lambda n \rceil}^{n} \binom{n}{k} \leq 2^{nh(\lambda)}$$

where $h(\lambda) = -[\lambda \log \lambda + (1 - \lambda)\log(1 - \lambda)]$.

13. A memoryless source of entropy 15 bits per sourceword is connected to a binary symmetric channel of error probability 0.1 and which can pass 10^3 binary digits per second through to the receiver.

At what rate can the source words be emitted if they are to be communicated accurately through the channel?

4

Error-correcting codes

4.1 The coding problem

Shannon's theorem assures us that good codes exist. The difficulty is
finding them. The aim of this chapter is to give some indication of the
different approaches to this problem.

First, a brief recap of notation and ideas. The situation is best
described in terms of Fig. 1.

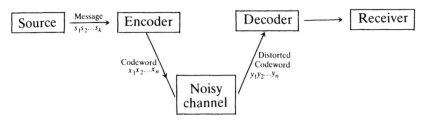

Fig. 1

The source will produce a *message* consisting of a sequence of
source symbols, and this message is to be transmitted to its intended
receiver across a noisy channel.

Without any real loss of generality, we assume that the channel has
the same alphabet Σ, of size q, for input and output. A *code* $\mathscr{C}$ over
Σ is a collection of sequences of symbols from Σ; the members of $\mathscr{C}$
are *codewords*. We assume that all codewords are of the same length.
Using such codes (known as *block* codes) makes decoding much
easier. If the codewords of $\mathscr{C}$ have length n and $|\Sigma| = q$, then the
code is described as a *q-ary code of length n* (*binary* when $q = 2$ and
ternary when $q = 3$).

We let $V_n(\Sigma)$ denote the set of all n-sequences of symbols from the
alphabet Σ, and call the elements of $V_n(\Sigma)$ *vectors* or *words*.
Sometimes, where it is more important to remind the reader that Σ
has q symbols we write $V_n(\Sigma)$ as $V_n(q)$, while as already defined, V_n
stands for the set of binary words of length n.

As in the binary case, the (*Hamming*) *distance* $d(x, y)$ between two vectors x and y is the number of places in which x and y differ; *minimum-distance* or *nearest-neighbour* decoding, means decoding a received vector y into a codeword c that is a minimum distance from y (where there is a choice, choose arbitrarily).

If $\mathscr{C}$ is a code, then the *minimum distance* of $\mathscr{C}$, written

$$d(\mathscr{C}) = \min d(c_i, c_j),$$

where the minimum is taken over all pairs of distinct codewords in $\mathscr{C}$. It is a key concept in evaluating a code: good codes have their codewords scattered so that their minimum distance is large.

The reason why the minimum distance of a code is such an important parameter is clear from the following easy theorem.

Theorem 1 *If a code has minimum distance d, the minimum-distance decoding scheme will correct up to $\frac{1}{2}(d-1)$ errors.*

Proof Take $e = \lfloor \frac{1}{2}(d-1) \rfloor$ and consider the *e-sphere* surrounding x. This is the set $S_e(x)$ given by

$$S_e(x) = \{y : d(x, y) \le e\}.$$

Because of the minimum-distance hypothesis, if x and z are distinct codewords, then

$$S_e(x) \cap S_e(z) = \varnothing.$$

Hence minimum-distance decoding will correct up to e errors. □

If a code has M codewords of length n and has minimum distance d, then it is called an (n, M, d)-code. For fixed n, the parameters M and d pull against each other, in the sense that increasing M tends to decrease d and vice versa.

We let $A_q(n, d)$ denote the maximum M such that there exists a q-ary (n, M, d) code. Clearly, by taking the code of all n vectors, we have

(1)
$$A_q(n, 1) = q^n,$$

but, for even relatively small values of q, n, and d, the quantity $A_q(n, d)$ is not known.

For example, in the binary case, it is an open problem to determine $A_2(10, 3)$ exactly. All that is presently known is that

(2)
$$72 \le A_2(10, 3) \le 79.$$

For $n = 11$, the corresponding result is

(3)
$$144 \le A_2(11, 3) \le 158.$$

Basically, the problem of determining $A_2(n, d)$ is a problem in finite geometry (albeit a pretty intractable one).

For example, in the case $n = 3$ (see Fig. 2), the set of points A, B, C, D is a $(3, 4, 2)$ code, and this shows that $A_2(3, 2) \ge 4$ (it is, in fact, equal to 4).

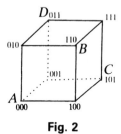

Fig. 2

A very useful device for reducing the amount of work involved in finding good codes is to use the notion of equivalence, which we now define.

EQUIVALENT CODES

Suppose that we have an (n, M, d)-code $\mathscr{C}$. The natural way to present it is by an $M \times n$ array whose rows are the distinct codewords.

Now suppose that π is any permutation of $(1, 2, \ldots, n)$ and that, for each codeword $c \in \mathscr{C}$, we apply the transformation

$$\pi : c \mapsto c'$$

defined by

$$c_i' = c_{\pi(i)} \quad (1 \le i \le n).$$

We call such a transformation a *positional permutation*.

In the same way, if π is any permutation of the symbols in Σ, we say that π induces a *symbol permutation* of $\mathscr{C}$ if, for some i, with $1 \le i \le n$, and for each codeword $c \in \mathscr{C}$, we transform by

$$c \mapsto c',$$

where c' is defined by

$$c_j' = c_j \quad (1 \le j \le n, \ j \ne i), \qquad c_i' = \pi(c_i).$$

If a code $\mathscr{C}'$ can be obtained from a code $\mathscr{C}$ by a sequence of positional or symbol permutations, then $\mathscr{C}'$ is called an *equivalent* code.

Example   Suppose that $\mathscr{C}$ is a code of length 5 over the alphabet $\Sigma = \{a, b, c\}$ having codewords c_1, c_2, c_3, c_4 as shown:

$$\mathscr{C} \equiv \begin{array}{c} c_1 \\ c_2 \\ c_3 \\ c_4 \end{array} \begin{bmatrix} a & b & c & a & c \\ b & a & b & a & b \\ b & c & c & b & a \\ c & b & a & c & a \end{bmatrix}.$$

The shift permutation $\{1 \mapsto 2,\ 2 \mapsto 3,\ 3 \mapsto 4,\ 4 \mapsto 5,\ 5 \mapsto 1\}$ induces a positional permutation which shows $\mathscr{C}$ is equivalent to

$$\mathscr{C}' = \begin{array}{c} c_1' \\ c_2'' \\ c_3'' \\ c_4' \end{array} \begin{bmatrix} c & a & b & c & a \\ b & b & a & b & a \\ a & b & c & c & b \\ a & c & b & a & c \end{bmatrix}.$$

Similarly the permutation $\{a \mapsto b,\ b \mapsto c,\ c \mapsto a\}$, applied to the first symbol of each word in $\mathscr{C}'$, induces the symbol permutation

$$\mathscr{C}'' = \begin{array}{c} c_1'' \\ c_2'' \\ c_3'' \\ c_4'' \end{array} \begin{bmatrix} a & a & b & c & a \\ c & b & a & b & a \\ b & b & c & c & b \\ b & c & b & a & c \end{bmatrix}. \qquad \square$$

The notion of equivalence is useful for the following reasons.

(4)   If $\mathscr{C}$ and $\mathscr{C}'$ are equivalent codes, then the set of distances between the codewords of $\mathscr{C}$ is identical with that between the codewords of $\mathscr{C}'$.

Proof   Show that both symbol and positonal permutations do not affect distances, hence no sequence of them does. $\square$

(5)   If $\mathscr{C}$ is any code of length n, and u is any n-vector over the same alphabet, then there exists a code $\mathscr{C}'$ that contains u and is equivalent to $\mathscr{C}$.

Proof   The first codeword c_1 can be transformed into u by at most n symbol permutations. $\square$

1. Prove that $A_2(3, 2) = 4$.

2. Show that, for any codewords x, y, z of the same length, the Hamming distance satisfies

$$d(x, y) + d(y, z) \geq d(x, z).$$

3. Prove that the r-sphere $S_r(x)$ surrounding the vector $x \in V_n(q)$ contains exactly

$$1 + \binom{n}{1}(q - 1) + \ldots + \binom{n}{r}(q - 1)^r$$

vectors.

4.2 The sphere-packing and Gilbert–Varshamov bounds; perfect codes

If is very easy to get what is known as a *sphere-packing* upper bound on $A_q(n, d)$. This states that, when d is odd, say $d = 2e + 1$, then

(1)
$$A_q(n, d) \sum_{k=0}^{e} \binom{n}{k}(q - 1)^k \leq q^n.$$

Proof Let $\mathscr{C}$ be a code with minimum distance $2e + 1$; then, if $S_e(x)$ denotes the e-sphere surrounding x, we have

$$S_e(x) \cap S_e(y) = \varnothing$$

for any pair of distinct codewords x and y. But a simple counting argument shows that the cardinality of $S_e(x)$ is given by

(2)
$$|S_e(x)| = \sum_{k=0}^{e} \binom{n}{k}(q - 1)^k.$$

The right-hand side of (1) is the total number of words of length n from an alphabet of q symbols.

Combining these two observations gives (1). □

An equally easy argument gives what is known as the Gilbert–Varshamov bound, which states

(3)
$$A_q(n, d) \sum_{i=0}^{d-1} \binom{n}{i}(q - 1)^i \geq q^n.$$

Proof Suppose that $\mathscr{C}$ is an (n, M, d) code with a maximum number of codewords. Then there can be no vector in $V_n(q) \setminus \mathscr{C}$ that is distant at least d from all the codewords of $\mathscr{C}$. In other words, the $(d - 1)$-

spheres surrounding the codewords must cover $V_n(q)$. But this is reduced to the condition (3) by using (2).   □

PERFECT CODES

The ideal situation, from the point of view of economy, is to have a code $\mathscr{C}$ over $V_n(q)$ such that, for some $t > 0$, the t-spheres around the codewords of $\mathscr{C}$ are disjoint but their union contains every vector in $V_n(q)$. Such a code is called *perfect*.

From its definition, it is clear that such a code (a) can correct up to t errors, and (b) cannot correct $t + 1$ errors by minimum-distance decoding.

Trivial consequences of the definition are the following.

(4)   A necessary condition for an (n, M, d)-code to be perfect is that d is odd.

(5)   An (n, M, d)-code is perfect if and only if d is odd and

(6)
$$M \sum_{i=0}^{\frac{1}{2}(d-1)} \binom{n}{i}(q-1)^i = q^n.$$

Example   Two obvious classes of perfect codes are the following:
(a) any code with exactly one codeword,
(b) any binary code with just the two words of odd length

$$00 \ldots 0 \quad \text{and} \quad 11 \ldots 1. \qquad \square$$

These codes are called the *trivial perfect codes*. We shall meet more interesting examples later.

Exercises   1. Show that
4.2
$$19 \leq A_2(10, 3) \leq 93.$$

2. Show that, for any positive integer q, the parameters

$$n = (q^r - 1)/(q - 1), \qquad M = q^{n-r}, \qquad d = 3,$$

where r is any integer ≥ 2, satisfy the conditions (6) for the parameters of a perfect code.

Note: Although these parameters satisfy the conditions (6) for any positive integer, it is conjectured that there exist perfect codes with these parameters only when q is a prime power.

4.3 Linear codes

Suppose $\mathscr{C}$ is a code which has a minimum distance $d = 2e + 1$, so that we can correct up to e errors by nearest-neighbour decoding. If $\mathscr{C}$ has very few members, this is practicable. However, if $|\mathscr{C}|$ is large, this can be very time consuming if it involves comparing the received vector y with each of a large number of codewords. One of the reasons for studying more highly structured codes (such as the linear codes of this section) is to try to avoid this problem.

Let us suppose that the size of our alphabet Σ is a prime power q(say), so that we can regard Σ as the set of distinct elements of the field F_q with q elements.

We can now let $V_n(q)$ be the vector space of dimension n over F_q, a typical member of which will be denoted by $x = (x_1, x_2, \ldots, x_n)$, which, for brevity, we will sometimes write as $x = x_1 \ldots x_n$ where $x_i \in F_q$.

A *linear code* $\mathscr{C}$ over Σ is defined to be any subspace of $V_n(q)$; if $\mathscr{C}$ is a k-dimensional subspace, then we call it an $[n, k]$-*code* or, when we know (or wish to specify) its minimum distance, we may call it an $[n, k, d]$-code.

Since any k-dimensional subspace over F_q has q^k members, we have:

(1) $\qquad\qquad$ An $[n, k, d]$-code over F_q is an (n, q^k, d)-code.

This highlights one of the obvious advantages of linear codes: they allow k codewords of length n to describe completely a code with q^k such codewords. The *space savings* can be enormous. This is because any subspace of dimension k is completely specified once k linearly independent vectors of the subspace are known.

Accordingly we define a *generator matrix* for a linear $[n, k]$-code $\mathscr{C}$ to be any $k \times n$ matrix whose rows constitute k linearly independent codewords of $\mathscr{C}$.

Now suppose G is a generator matrix of $\mathscr{C}$ and G' is any other matrix obtained from G by any finite sequence of operations of the following types:

(a) permuting rows,
(b) multiplying a row by a nonzero scalar,
(c) adding to a row a scalar multiple of another row,
(d) permuting columns,
(e) multiplying any column by a nonzero scalar.

Then it is a routine exercise in linear algebra to prove the following.

Lemma 1   *G' is the generator matrix of a code $\mathscr{C}'$ which is equivalent to $\mathscr{C}$.*

Proof   Given the generator matrix G of the linear code $\mathscr{C}$, we have to show that each of the operations (a)–(e) give a matrix G' which is the generator matrix of an equivalent code $\mathscr{C}'$. Now, (a) is trivial and, since (b) and (c) are linear operations on the rows, each of these operations gives a code $\mathscr{C}'$ which is not only equivalent but identical with $\mathscr{C}$. Let C be the full code matrix of $\mathscr{C}$. Again, permuting the columns of G is equivalent to permuting each column of the full code matrix C.

Multiplying a column by a nonzero scalar is in effect permuting the elements of Σ in that column, since we are working with a linear code. This completes the proof.   □

But, again using routine methods of linear algebra, it is easy to prove the following lemma.

Lemma 2   *Let G be any $k \times n$ matrix whose rows are linearly independent; then, by applying a sequence of operations of type (a)–(e) to G, it is possible to transform G to a matrix of type $[I_k, A]$, where I_k is the $k \times k$ identity matrix.*

Proof   Similar to that of Lemma 1 and left as an exercise.   □

Combining these two lemmas in the obvious way, we get a proof of the next theorem.

Theorem 1   *If $\mathscr{C}$ is any linear $[n, k]$-code, then there exists an equivalent code $\mathscr{C}'$ with generator matrix $[I_k, A]$, where I_k is the $k \times k$ identity matrix.*

Because of this theorem, we will usually think of generator matrices of codes as being in this standardized form.

Another useful property of linear codes is that their minimum distance is more easily found than is the case with general codes. We define the *weight* $w(x)$ of a vector to be its number of nonzero coordinates. Then we have the next result.

Theorem 2   *The minimum distance of a linear code $\mathscr{C}$ is the minimum weight of a nonzero vector in $\mathscr{C}$.*

Proof   Let d be the minimum distance of $\mathscr{C}$ and suppose x and y are codewords with $d(x, y) = d$.

Since $\mathscr{C}$ is a linear subspace, the vector $x - y$ is also a codeword of $\mathscr{C}$. But $w(x - y) = d$, showing that the minimum weight is at most d. It clearly cannot be strictly less than d, because, if $z \neq 0$ were such that $w(z) < d$, we would have the contradiction

$$w(z) = d(z, \mathbf{0}) < d. \qquad \square$$

Exercises 4.3

1. Consider the binary linear code $\mathscr{C}$ which has generator matrix given by

$$\begin{bmatrix} 1 & 0 & 1 & 1 & 1 \\ 0 & 1 & 1 & 0 & 1 \\ 1 & 1 & 0 & 0 & 0 \end{bmatrix}.$$

Find a matrix A such that the code with generator matrix $[I_3, A]$ is equivalent to $\mathscr{C}$.

2. How many codewords does the code $\mathscr{C}$ of exercise 1 have? What is its minimum distance?

3. Assume that, over a given field, there exists an $[n_1, k, d_1]$-code and an $[n_2, k, d_2]$-code. If G_1 and G_2 are the respective generator matrices, consider $[G_1, G_2]$. What are the parameters of the code it generates?

4.4 Using linear codes

Suppose that $\mathscr{C}$ is a linear $[n, k]$-code over $F_q = \Sigma$ and that it has generator matrix G, defined as

$$G = \begin{bmatrix} r_1 \\ r_2 \\ \vdots \\ r_k \end{bmatrix} = [I_k, A],$$

where the r_i are n-vectors over F_q and A is a $k \times (n - k)$ matrix. The codewords of $\mathscr{C}$ are all of the q^k vectors of length n of the form

$$\sum_{i=1}^{k} a_i r_i, \qquad a_i \in F_q.$$

The basic idea of encoding is as follows. If the message sequence is $s = (s_1, \ldots, s_k)$, we encode s by the codeword $c = (c_1, \ldots, c_n)$ where the c_i are given by the rule

(1)
$$c_i = s_i \quad (1 \leq i \leq k),$$

(2)
$$\begin{bmatrix} c_{k+1} \\ \vdots \\ c_n \end{bmatrix} = \begin{bmatrix} s_1 \\ \vdots \\ s_k \end{bmatrix} A.$$

Example Suppose that a code over F_3 (the field of integers modulo 3) has generator matrix

$$G = \begin{bmatrix} 1 & 0 & 0 & 1 & 2 & 0 \\ 0 & 1 & 0 & 0 & 1 & 1 \\ 0 & 0 & 1 & 2 & 0 & 1 \end{bmatrix}.$$

If the input message stream from the source is

$$102101210122\dots\dots$$

we first break it up into blocks of length 3 to get

$$102 \mid 101 \mid 210 \mid 122 \mid \dots\dots$$

and then encode the source words as

$$102 \mapsto r_1 + 2r_3 = 102222, \qquad 101 \mapsto r_1 + r_3 = 101021,$$
$$210 \mapsto 2r_1 + r_3 = 210221, \qquad 122 \mapsto r_1 + 2r_2 + 2r_3 = 122211,$$

where addition is, of course, mod 3. Hence the message stream passed to the channel by the encoder is the sequence

$$102\ 222\ 101\ 021\ 210\ 221\ 122\ 211\dots\dots$$

Clearly, by doubling the length of the message, we are halving the rate of transmission. Hopefully, we are increasing the reliability! □

Going back to equations (1) and (2), we see that they can be rewritten in the form

$$\begin{bmatrix} s_1 \\ \vdots \\ s_k \end{bmatrix} A = \begin{bmatrix} c_1 \\ \vdots \\ c_k \end{bmatrix} A = \begin{bmatrix} c_{k+1} \\ \vdots \\ c_n \end{bmatrix},$$

so that they are equivalent to

$$[-A^\mathsf{T}, I_{n-k}]c^\mathsf{T} = 0.$$

The matrix

(3) $$H = [-A^\mathsf{T}, I_{n-k}]$$

is called the *parity-check matrix* of the code. What we have shown above is that:

(4)   A vector z is a codeword of $\mathscr{C}$ if and only if $Hz^\mathsf{T} = 0$.

The parity check matrix H clearly defines the code just as well as the generator matrix G, and it owes its name to the fact that what we are doing is adding on some check digits to detect/correct errors in our codewords. For that reason, in an $[n, k]$-code, the first k digits of a codeword are often called the *message* digits and the remaining $n - k$ are the *check* or *parity-check* digits.

Example 2

(*Example 1 revisited*) Consider the generator matrix G of Example 1. By (3), the parity-check matrix H is given by

$$H = \begin{bmatrix} -1 & 0 & -2 & 1 & 0 & 0 \\ -2 & -1 & 0 & 0 & 1 & 0 \\ 0 & -1 & -1 & 0 & 0 & 1 \end{bmatrix}$$

$$= \begin{bmatrix} 2 & 0 & 1 & 1 & 0 & 0 \\ 1 & 2 & 0 & 0 & 1 & 0 \\ 0 & 2 & 2 & 0 & 0 & 1 \end{bmatrix},$$

since we are working in the field F_3 of integers mod 3. The codeword 102 222 consisting of

$$\text{message digits } 102$$

and

$$\text{check digits } 222$$

must satisfy the equations

$$Hc^\mathsf{T} = 0,$$

which reduce to

(5)
$$2c_1 + c_3 + c_4 = 0, \qquad c_1 + 2c_2 + c_5 = 0, \qquad 2c_2 + 2c_3 + c_6 = 0,$$

and so on for each of the other codewords.

The crude idea of the error correction method can be seen in this example. Suppose that our received codeword failed to satisfy the first and third of the parity check equations (5). Then we would deduce that the message digit in error is c_3, since this is the only digit occurring in both equations. □

Exercises 4.4

1. A binary code is described by its parity-check matrix

$$H = \begin{bmatrix} 1 & 1 & 1 & 0 & 1 & 0 & 0 \\ 1 & 1 & 0 & 1 & 0 & 1 & 0 \\ 1 & 0 & 1 & 1 & 0 & 0 & 1 \end{bmatrix}.$$

How would you encode the message $m = 0000\,1101$?

2. Using the same code how would you decode the received vector $0000111\,0001110$?

3. Find the parameters n, M, d of the code whose parity-check matrix is given above.

4.5 Minimum-distance decoding for linear codes

Consider the decoding problem for linear codes. If $\mathscr{C}$ is an $[n, k]$-code over the alphabet $\Sigma = F_q$, then $\mathscr{C}$ contains q^k codewords of length n and the number of possible received vectors is q^n.

A look-up table containing for each possible received vector a 'nearest' codeword would consume a great deal of space, even for moderate values of k and n. One of the major advantages of using linear codes is that there is an elegant mathematical way of avoiding this problem.

We describe this, working only in the binary case. The extension to other alphabets of prime-power size is fairly straightforward, once the basic idea is understood.

So, suppose $\mathscr{C}$ is an $[n, k]$-binary code. Since $\mathscr{C}$ is a subspace of the vector space V_n of binary n-vectors, $\mathscr{C}$ must be a subgroup of the additive group determined by V_n.

Thus, $\mathscr{C}$ determines a collection of cosets of V_n, where (we remind the reader) an arbitrary vector a of V_n determines a unique coset $a + \mathscr{C}$, defined by, $a + \mathscr{C}$ consists of all vectors b of V_n that can be written in the form

$$b = a + c$$

for some $c \in \mathscr{C}$.

Now suppose y is a received vector when some codeword is transmitted through the channel. We say a vector e is a *possible error vector* of y if there is some codeword $c \in \mathscr{C}$ such that

$$y - c = e.$$

The interpretation is obvious: a vector e is an error vector associated with a received vector if it could represent a possible pattern of errors in transmission. A key observation, but a trivial one, is the following.

Lemma *If y is the received vector, then the set of possible error vectors is the coset of $\mathscr{C}$ that contains the vector y.*

Proof If y is received, then e is an error vector of y if and only if there exists a codeword $c \in \mathscr{C}$ such that $e = y - c$. But $\mathscr{C}$ is a subspace; so, if $c \in \mathscr{C}$, then $-c \in \mathscr{C}$ and thus $e = y + c'$ (with $c' = -c$) and $e \in y + \mathscr{C}$. □

Now what is minimum-distance decoding? It is merely the problem of finding an error vector of minimum weight. Hence if we know, for each coset, a member of the coset of minimum weight, then we have the basis of a minimum-distance decoding rule. Accordingly, for each coset, we call a vector a *coset leader* if that coset contains no other vector having smaller weight. We emphasize that a coset leader is not uniquely determined.

Algorithm *Step 1*: On receiving y, find a coset leader z_0 of the coset determined
I   by y.
             Step 2: Decode y as the codeword $y - z_0$.

As it stands, Step 1 can be time-consuming. We speed it up using the following property of linear codes.

(1)   Two vectors y_1 and y_2 are in the same coset with respect to $\mathscr{C}$ if and only if
$$Hy_1^T = Hy_2^T$$

Proof   The vectors y_1 and y_2 are in the same coset if and only if there exists some codeword c such that
$$y_1 = y_2 + c.$$
But $Hc^T = 0$ by (4.4).   □

Define the *syndrome* of the coset $a + \mathscr{C}$ to be the vector Ha^T; by (1) it is well defined. Thus, associated with each coset, we have a syndrome and a coset leader. Hence, if we have a *look-up table*, which gives the corresponding coset leader for each syndrome, we can speed up the above algorithm by replacing it by the following one.

Algorithm *Step 1(a)*: On receiving y, calculate its syndrome Hy^T.
II (Semi- *Step 1(b)*: From the above look-up table, read off the corresponding
efficient) coset leader z_0.
             Step 2: Decode y as the vector $y - z_0$.

Before analysing Algorithm II, we first prove the following result.

Theorem *Algorithm II is a minimum-distance decoding scheme for the linear*
I   *code $\mathscr{C}$.*

Proof   First note that any received vector y is decoded as a codeword. This is because y and z_0 are in the same coset; hence $y - z_0 \in \mathscr{C}$.
             Suppose there is a codeword c with
$$d(y, y - z_0) > d(y, c).$$

Then, equivalently,

$$d(z_0, \mathbf{0}) > d(y - c, \mathbf{0}).$$

But this means that the weight $w(z_0) > w(y - c)$. But

$$H(y - c)^T = Hy^T - Hc^T = Hy^T$$

since c is a codeword; so $y - c$ has the same syndrome as y and, by (1), belongs to the same coset as y and has weight strictly less than that of the coset leader z_0, which is a contradiction.   □

The disadvantages of this coding method—hinted at in our description of it as 'semi-efficient'—are best seen after considering an example.

Example Suppose that $\mathscr{C}$ is a binary code whose generator matrix G is given by

$$G = \begin{bmatrix} 1 & 0 & 1 & 0 \\ 0 & 1 & 1 & 1 \end{bmatrix};$$

then its parity-check matrix is

$$H = \begin{bmatrix} 1 & 1 & 1 & 0 \\ 0 & 1 & 0 & 1 \end{bmatrix}.$$

The codewords of $\mathscr{C}$ are

$$0000, 1010, 0111, 1101.$$

The syndrome look-up table is

Syndrome	Coset Leader = (Corrector)
00	0000
10	0010
01	0001
11	0100.

Hence suppose we receive the vector $y = 1111$. Its syndrome is $Hy^T = 10$. The corresponding corrector or coset leader is 0010 and hence 1111 is decoded as 1101.

Note that this look-up table is not unique: the corrector of 10 could well have been taken to be 1000. In this case, 1111 would have been decoded as 0111.   □

In the general case of an $[n, k]$-code, there will be $|V_n| / |\mathscr{C}| = 2^{n-k}$ different cosets; hence the look-up table used in Step 1(b) will have 2^{n-k} different entries.

Searching this for the coset leader is laborious when n and k are large. Nevertheless, the overall advantage of the method means it is widely used.

Exercises 4.5

1. A binary code has parity-check matrix H given by

$$H = \begin{bmatrix} 1 & 1 & 0 & 0 & 1 & 0 & 0 \\ 0 & 0 & 1 & 1 & 0 & 1 & 0 \\ 1 & 1 & 1 & 1 & 0 & 0 & 1 \end{bmatrix}.$$

Construct the syndrome look-up table, and hence decode the following received words:

(a) 1111111, (b) 1101011,

(c) 0110111, (d) 0111000.

4.6 Binary Hamming codes

To illustrate the above techniques consider the following example. Restrict attention to the binary case; take r to be any positive integer; and let $n = 2^r - 1$. Take H to be the $r \times (2^r - 1)$ matrix whose columns are the distinct nonzero vectors of V_r. Then H is the parity-check matrix of a binary $[n, k]$-code, where

$$n = 2^r - 1, \qquad k = n - r.$$

This code is called the $[n, k]$ *Hamming code*.

The key property of Hamming codes is summed up in the following theorem.

Theorem 1

Any Hamming code is a perfect single-error-correcting code.

Proof

We first show that the minimum distance of such a Hamming code $\mathscr{C}$ is at least 3. Because $\mathscr{C}$ is a linear code, we know from Theorem 3.2 that the minimum distance $d(\mathscr{C})$ equals the minimum weight of a vector in $\mathscr{C}$.

Suppose that $\mathscr{C}$ has a codeword u of weight 1 with a nonzero entry in the ith place. Then

$$Hu^{\mathsf{T}} = 0 \quad \Rightarrow \quad \text{the } i\text{th column of } H \text{ is zero},$$

which is obviously untrue.

Suppose $\mathscr{C}$ has a codeword v of weight 2, with nonzero entries in the ith and jth places. Then

$$Hv^{\mathsf{T}} = 0 \quad \Rightarrow \quad h_i + h_j = 0,$$

where h_i denotes the ith column of H. But this means H has 2 identical columns which is nonsense.

Thus $d(\mathscr{C}) \geq 3$. (In fact, just by finding a codeword of weight 3, we can see that $d(\mathscr{C}) = 3$, but this is irrelevant).

To show that $\mathscr{C}$ is perfect, just notice that the 1-sphere surrounding any codeword $x \in \mathscr{C}$ will contain $1 + n = 2^r$ vectors. Since $\mathscr{C}$ contains $2^k = 2^{n-r}$ codewords, the union of these 1-spheres is the complete set of 2^n vectors in V_n, which completes the proof.   □

An important consequence of the Hamming code being perfect is that:

(1)   For an $[n, k]$-Hamming code, the coset leaders are exactly the vectors of V_n of weight ≤ 1.

This leads by a 'trick' to the following elegant decoding algorithm for Hamming codes. First note:

(2)   The columns of H can be rearranged so that the jth column of H is just the binary representation of the decimal integer j.

When a vector y is received, calculate its syndrome Hy^T and suppose it represents the decimal integer j. Assuming only a single error, the maximum-likelihood (=minimum distance) decoding scheme gives:

(i) If $j = Hy^T = 0$, then assume no error and y is a codeword.

(ii) If $j = Hy^T \neq 0$, then decode y by assuming an error in the jth position of y.

Example   The $[7, 4]$-Hamming code has parity-check matrix

$$H = \begin{bmatrix} 0 & 0 & 0 & 1 & 1 & 1 & 1 \\ 0 & 1 & 1 & 0 & 0 & 1 & 1 \\ 1 & 0 & 1 & 0 & 1 & 0 & 1 \end{bmatrix}.$$

Suppose we receive vector $y = (1\,0\,1\,0\,1\,1\,0)$, so that $Hy^T = (0\,0\,1)$. Then, on the assumption that there is not more than a single error, we assume the error occurs in the first place and hence decode y as y^* where

$$y^* = (0\,0\,1\,0\,1\,1\,0).$$ □

Exercises   1. Write down the parity-check matrix of a binary $[15, 11, 3]$-code. How
4.6   would it decode the received vectors

(a) $(100\,000\,000\,000\,000)$ (b) $(111\,111\,111\,111\,111)$?

4.7 Cyclic codes

We now discuss an important subclass of linear codes. A code $\mathscr{C}$ is *cyclic* if the following conditions hold.
(a) $\mathscr{C}$ is linear;
(b) if $w = (w_1, \ldots, w_n)$ belongs to $\mathscr{C}$, then so does $w' = (w_n, w_1, w_2, \ldots, w_{n-1})$.
These codes have attractive algebraic properties and are also easily constructed using linear shift registers as we describe in Chapter 8.

We work solely in the binary case and, throughout this section, identify a vector

$$w = (w_1, \ldots, w_n)$$

with the polynomial

$$w(x) = w_1 + w_2 x + w_3 x^2 + \ldots + w_n x^{n-1}.$$

Further, our arithmetic will be in the ring R_n of binary polynomials modulo $x^n - 1$. That is, R_n consists of polynomials of degree $\leq n - 1$, with coefficients 0 and 1, and with the rules of addition and multiplication given by

$$a(x) + b(x) = \sum_{i=0}^{n-1} (a_i + b_i)x^i,$$

$$a(x)b(x) = a(x)b(x) \pmod{x^n - 1}.$$

The fundamental observation is that a shift in a codeword w corresponds to multiplying the corresponding polymonial by x in the ring R_n.
From this it is easy to see:

(1) If $w(x)$ is the polynomial representation of a codeword in $\mathscr{C}$, then so is $w(x)f(x)$ for any polynomial f of degree $\leq n - 1$.

Proof Since $w(x) \in \mathscr{C}$, the polynomial $xw(x)$ is associated with a 1-shift, and hence belongs to $\mathscr{C}$. Iterating, we note

$$x^k w(x) \in \mathscr{C}$$

for any integer k. But $\mathscr{C}$ is linear, so any linear combination of codewords is a codeword; thus $f(x)w(x) \in \mathscr{C}$ for any polynomial f. $\square$

(2) Let $g(x)$ be a nonzero polynomial of minimum degree in $\mathscr{C}$; then $g(x)$ *generates* $\mathscr{C}$ in the sense that any codeword $w(x) \in \mathscr{C}$ can be

written in the form

$$w(x) = f(x)g(x)$$

for a suitable polynomial f.

Proof   Assume that there is some $w(x)$ that cannot be so written. Then $w(x)$, when divided by $g(x)$, leaves a nonzero remainder $r(x)$ of degree strictly less than $g(x)$. Write

$$w(x) = q(x)g(x) + r(x).$$

Since $w(x)$ and $q(x)g(x)$ are codewords, then, because $\mathscr{C}$ is linear, we have $r(x) \in \mathscr{C}$. This contradicts the minimum degree of g.   □

Accordingly, we call $g(x)$ the *generator polynomial* of the code $\mathscr{C}$. It clearly gives a very compact representation of a code.

Note   Those familiar with ring theory will recognize a cyclic code as just an ideal in the polynomial ring, and (2) then follows because every such ideal is principal.

Example   Suppose we take $n = 3$, so that the multiplication is modulo the polynomial $x^3 - 1$. It is easy to check that the code $\mathscr{C}$ given by

$$\mathscr{C} = \{0, 1+x, x+x^2, 1+x^2\}$$

is cyclic and generated by $1+x$. A standard representation of $\mathscr{C}$ consists of the vectors

$$\{(000), (110), (011), (101)\}.$$ □

Moreover, it is easy to obtain the usual generator matrix G of a cyclic code from its generator polynomial. This is given by the rule:

(3)   If $\mathscr{C}$ is a cyclic code of length n with generator polynomial $g(x) = g_1 + g_2 x + \ldots + g_k x^{k-1}$, then its generator matrix is given by the $(n - k + 1)$ by n matrix

$$G = \begin{bmatrix} g_1 & g_2 & g_3 & \cdots & g_k & 0 & 0 & \cdots & 0 \\ 0 & g_1 & g_2 & \cdots & g_{k-1} & g_k & 0 & \cdots & 0 \\ 0 & 0 & g_1 & \cdots & g_{k-2} & g_{k-1} & g_k & \cdots & 0 \\ \vdots & \vdots & & \cdots & & \vdots & \vdots & \vdots & & \vdots \\ 0 & 0 & \cdots & \cdots & & \cdots & \cdots & \cdots & g_k \end{bmatrix}$$

Proof   The rows of G are easily seen to be linearly independent. Hence, if we show that any codeword can be represented by a linear combination of these rows, we are home. Now by definition, a vector $c = (c_0 c_1 \ldots c_{n-1})$ is a codeword iff the corresponding polynomial

$$c(x) = c_0 + c_1 x + \ldots + c_{n-1} x^{n-1}$$

is of the form

$$c(x) = g(x)f(x) \quad (\mathrm{mod}(x^n - 1))$$

for some polynomial f which can be taken to be of degree $\leq n - 1$. But this means (in the obvious notation) that

$$c(x) = g(x)(f_0 + f_1 x + f_2 x^2 + \ldots + f_{n-1} x^{n-1})$$

$$= \sum_{i=0}^{n-1} f_i x^i g(x) \quad (\mathrm{mod}(x^n - 1));$$

and this, in turn, is exactly the statement that

$$c = f_0 g + f_1 g^{(1)} + \ldots + f_{n-1} g^{(n-1)},$$

where $g = (g_1, g_2, \ldots, g_n)$ and $g^{(k)}$ denotes the cyclic shift of g by k places. $\square$

Finally, it is natural to ask whether any polynomial will act as a generator polynomial of some cyclic code. To see that this is not true, note:

(4)   If $g(x)$ is the generator polynomial of the cyclic code $\mathscr{C}$ of length n, then g divides $x^n - 1$.

Proof   If not, we can write

$$x^n - 1 = g(x)q(x) + r(x),$$

where r is a nonzero polynomial with lower degree than g. Since $q(x)g(x) \in \mathscr{C}$ and $r = -qg$ in this ring, the linearity of $\mathscr{C}$ implies $r \in \mathscr{C}$, and thus contradicts the definition of g as the polynomial of minimum degree in $\mathscr{C}$. $\square$

Now consider the easily checked statement:

(5)   Given any polynomial p of degree $< n$, the set of polynomials qp (modulo $x^n - 1$), where q runs through the polynomials of degree less than n, forms a cyclic code of length n.

Combining (4) and (5), we can sum up the situation as follows:

(6)   The cyclic codes of length n are obtained by taking the divisors of

$x^n - 1$ and forming the cyclic codes having these polynomials as their generators.

Thus, effectively, we have an automatic and relatively easy way of constructing all cyclic codes of a given length.

We close this elementary introduction to cyclic codes with a result completely analogous to the 'parity-check matrix' approach of general linear codes.

(7)   If $\mathscr{C}$ is a binary cyclic code of length n and with generator polynomial g of degree k, then a polynomial p of degree less than n is a codeword if and only if

$$p(x)h(x) = 0,$$

where h is the polynomial of degree $n - k$ satisfying

$$g(x)h(x) = x^n - 1.$$

The polynomial h is called the *check polynomial* of the code $\mathscr{C}$.

Proof   If $c(x)$ is a codeword, then we know

$$c(x) = f(x)g(x)$$

for some polynomial f. Hence, since $gh = 0$, we have $ch = 0$.

Conversely, suppose that p is a nonzero polynomial satisfying $p(x)h(x) = 0$. Then p must have degree at least k. Thus, if p is not a codeword, we know from (2) that it is not divisible by g and so there exists a polynomial r of degree strictly less than that of g, with

$$p(x) = q(x)g(x) + r(x).$$

Since $ph = 0$ and $qgh = 0$, we must have $rh = 0$. But, since the degree of r is strictly less than that of g, the condition $rh = 0$ is impossible unless $r = 0$.   □

Exercises
4.7

1. Show that there are exactly four binary cyclic codes of length 3 and find their generator polynomial in each case.
2. Show that the Hamming code with parameters $n = 7$, $k = 4$, and $d = 3$ is a cyclic code with check polynomial $x^4 + x^2 + x + 1$. What is its generator polynomial?

4.8   The Mariner code; Reed–Muller codes

In this section, we give two further examples of the way classical modern algebra has been used to construct and develop classes of codes.

HADAMARD CODES

The code used by the 1969 space probe Mariner in order to transmit pictures from space was one of the class of codes constructed as follows.

A *Hadamard matrix* is an $n \times n$ matrix H whose entries are either $+1$ or -1, such that

$$HH^\mathsf{T} = nI_n.$$

where I_n is the $n \times n$ identity matrix.

If A and B are square matrices of sizes m and n respectively, define their *Kronecker product* to be the $mn \times mn$ matrix

$$A \otimes B = \begin{bmatrix} A_{11}B & A_{12}B & \dots & A_{1m}B \\ \vdots & \vdots & & \vdots \\ A_{m1}B & A_{m2}B & \dots & A_{mm}B \end{bmatrix}.$$

Direct manipulation suffices to prove the following result.

Lemma 1   *If H_1 and H_2 are Hadamard matrices, then so is their Kronecker product $H_1 \otimes H_2$.*

Hence if we start with the smallest nontrivial Hadamard matrix

$$H_2 = \begin{bmatrix} 1 & 1 \\ 1 & -1 \end{bmatrix},$$

then we can successively iterate the Kronecker product to get a sequence of Hadamard matrices of (rapidly) increasing size.

To use these in coding, suppose H is an $n \times n$ Hadamard matrix, where n is even. Define A to be the $2n \times n$ matrix

$$A = \begin{bmatrix} H \\ -H \end{bmatrix}.$$

Then define M to be the matrix obtained from A by replacing every entry equal to -1 by 0. Almost by definition this implies:

(1)   If x and y are two distinct rows of M, then $d(x, y)$ is equal to $\frac{1}{2}n$. This gives:

(2)   The rows of M form a binary $(n, 2n, \frac{1}{2}n)$-code.

Doing this for the five-fold Kronecker product of H_2 above, so that $n = 32$, gives the code used by Mariner.

Codes made by this construction in general are called *Hadamard codes*.

REED–MULLER CODES

This is a class of practically important codes which derive their name from their discoverers I. S. Reed and D. E. Muller (c. 1954). In order to describe these codes we need first to present an easy way of constructing new codes from old.

Lemma 2 *Given an (n, M_1, d_1) binary code $\mathscr{C}_1$ and another (n, M_2, d_2) binary code $\mathscr{C}_2$ we define a third binary code $\mathscr{C}_3 = \mathscr{C}_1 * \mathscr{C}_2$ by*

$$\mathscr{C}_3 = \{(\boldsymbol{u}, \boldsymbol{u} + \boldsymbol{v}) : \boldsymbol{u} \in \mathscr{C}_1, \boldsymbol{v} \in \mathscr{C}_2\}.$$

Then $\mathscr{C}_3$ is an $(2n, M_1 M_2, d_3)$-code where

(3)
$$d_3 = \min\{2d_1, d_2\}.$$

Proof Straightforward and left to the reader. □

We are now in a position to define recursively the binary rth-order Reed–Muller code $\mathscr{C}(r, m)$ by the rules:

For any positive integer m and r, with $0 \le r \le m$, we define $\mathscr{C}(r, m)$ to be a code with length $n = 2^m$ such that

$$\mathscr{C}(0, m) = \{\boldsymbol{0}, \boldsymbol{1}\},$$

where $\boldsymbol{0} = (00\ldots 0)$ and $\boldsymbol{1} = (11\ldots 1)$, $\mathscr{C}(m, m)$ is the set of all vectors of length 2^m, and

$$\mathscr{C}(r + 1, m + 1) = \mathscr{C}(r + 1, m) * \mathscr{C}(r, m).$$

Thus we can build up the codes as follows.

$m = 1$	$\mathscr{C}(0, 1) = \{00, 11\}$,
	$\mathscr{C}(1, 1) = \{00, 10, 01, 11\}$,
$m = 2$	$\mathscr{C}(0, 2) = \{0000, 1111\}$,
	$\mathscr{C}(1, 2) = \mathscr{C}(1, 1) * \mathscr{C}(0, 1)$,

so that $\mathscr{C}(1, 2)$ has codewords

0000, 0011, 1010, 1001, 0101, 0110, 1111, 1100,

and so on.

Using Lemma 2, it is now easy to prove the next theorem.

Theorem 1 *For any positive integer m and any r, with $0 \le r \le m$, the Reed–Muller code $\mathscr{C}(r, m)$ is an (n_r, M_r, d_r) binary code, with*

(a)
$$M_r = 2^a,$$

where

$$a = 1 + \binom{m}{1} + \ldots + \binom{m}{r},$$

(b) $n_r = 2^m,$

(c) $d_r = 2^{m-r}.$

Proof   The proof is a straightforward consequence of Lemma 2, using the identity

$$\binom{m+1}{r+1} = \binom{m}{r+1} + \binom{m}{r}$$

to prove (a) being the only nontrivial observation.   □

Exercises   1. Prove that the Kronecker product of Hadamard matrices is a Hadamard
4.8   matrix.
            2. Another way of thinking about Reed–Muller codes is as follows.
                Let v_0 be the vector whose 2^m components are all 1's and let v_1,
                $v_2, \ldots, v_m$ be the rows of a matrix which has all possible m-tuples
                as columns. Now define the 'product' of two vectors

$$a = (a_1, \ldots, a_n), \qquad b = (b_1, \ldots, b_n)$$

                by

$$ab = (a_1 b_1, \ldots, a_n b_n).$$

                The rth-order Reed–Muller code is formed by using as a basis the vectors
                $v_0, v_1, \ldots, v_m$ and all vectors obtained from these by taking products of
                r or fewer of these vectors.
                Illustrate this construction by finding the generator matrix of $\mathscr{C}(2, 4)$
                and verifying its parameters satisfy the conditions (a)–(c) of Theorem 1.

4.9 Conclusion

All the codes discussed in this chapter have been block codes, that is,
their codewords have the same length. Important classes of codes do
not have this property; in particular *convolutional codes* have
become increasingly useful in applications.

A convolution encoder can be regarded as an encoding system such
that, at each time instant, k information digits enter the encoder.
Each information digit remains in the encoder for a time units and
may affect each output during that time. In other words the encoder
has memory a. For further details of these codes we refer to the

books of Blahut (1983), Gallager (1969), McEliece (1977), and van Lint (1982).

The block codes that we have studied have almost all been linear codes. The case for concentrating on linear codes is:

(a) they tend to be the bricks from which bigger codes are built and are easy to deal with mathematically;

(b) they have relatively fast encoding and decoding procedures—this is important since, at the rate at which digits travel nowadays, extremely complex encoding and decoding would make an otherwise good code practically useless.

(c) Elias (1955) gave an attractive extension of Shannon's noisy coding theorem by showing that essentially the same proof works for linear codes. This may be stated as follows.

Theorem 1 *For $\varepsilon > 0$ and $R < C(p)$, provided N is sufficiently large, there exists a binary linear code $\mathscr{C}$ having 2^{RN} words of length N such that, on the binary symmetric channel with bit error probability p, the error probability is less than ε.*

Thus, if they could be found, good codes exist within the class of linear codes. The search for these good algebraic codes has led to a rich and beautiful theory of which the above has been just a glimpse. There are several superb texts devoted entirely to the subject, in particular MacWilliams and Sloane (1977), van Lint (1982), or (at a more elementary level) Hill (1986).

PROBLEMS 4

1. Show how you would detect single errors in a code over any alphabet by adding a single check digit.
2. Estimate how long it would take to find $A_2(n, d)$ by using exhaustive search of all possibilities on a computer with a speed of one operation per microsecond. Give an estimate in years of how long it would take to find the exact value of $A_2(10, 3)$ by brute-force search.
3. Prove that $A_2(5, 3) = 4$ and show that there is a unique (up to equivalence) $(5, 4, 3)$-code over $\{0, 1\}$.
4. Prove that $A_2(n, d) \leq 2A_2(n - 1, d)$.
5. Prove that, in the binary alphabet, if there exists an $(n, M, 2k)$-code, then there exists a code with the same parameters but with all its codewords of even weight.
6. The parity-check matrix H of a binary code is given by

$$H = \begin{bmatrix} 1 & 1 & 0 & 0 & 1 & 0 & 0 \\ 0 & 0 & 1 & 1 & 0 & 1 & 0 \\ 1 & 1 & 1 & 1 & 0 & 0 & 1 \end{bmatrix}.$$

If any symbol has probability p of being incorrectly transmitted, what is the probability that a codeword is correctly received after transmission and decoding using the above code?

7. Prove that, if H is the parity-check matrix of a code of length n, then the code has minimum distance d if and only if every $d - 1$ columns of H are linearly independent but some set of d columns is linearly dependent.

8. (i) Use the previous problem to show that, if $\mathscr{C}$ is an $[n, k, d]$-code, then $d \le n - k + 1$. (ii) Prove that, for all positive integers n, d, and q,

$$A_q(n, d) \le q^{n-d+1}.$$

(This is called the Singleton bound (Singleton, 1964) though it appears to go back to Komamiya (1953).)

9. Prove that, if $\mathscr{C}$ is a binary nontrivial perfect code of minimum distance 7 and word length n, then n must equal 23.

10. If d is odd, prove that a binary (n, M, d)-code exists if and only if a binary $(n + 1, M, d + 1)$-code exists.

11. Show that, in any binary linear code, either all the codewords have even weight or half have even weight and half have odd weight.

12. Show that the ternary code with generator matrix

$$\begin{bmatrix} 1 & 0 & 1 & 1 \\ 0 & 1 & 1 & 2 \end{bmatrix}$$

is a perfect code.

13. A perfect t-error correcting binary code $\mathscr{C}$ of word length n is used with a binary symmetric channel of error probability p. What is the probability that, when a codeword is sent, it is incorrectly received?

14. Prove that, if a Hadamard matrix of order n exists, then n must be 1,2, or a multiple of 4.

    Note: It is conjectured that Hadamard matrices exist whenever n is a multiple of 4. At present (1986) the smallest such n for which a Hadamard matrix has not been constructed is 268.

15. $\mathscr{C}$ is a binary cyclic code. Prove that, if $\mathscr{C}$ does not contain $11 \ldots 1$, then all the codewords of $\mathscr{C}$ have even weight.

16. If $\mathscr{C}$ is a binary $[n, k]$-code, then its *dual* $\mathscr{C}^\perp$ is defined by

$$\mathscr{C}^\perp = \{y \in V_n : \langle y, x \rangle = 0 \; \forall \, x \in \mathscr{C}\},$$

where $\langle x, y \rangle = x_1 y_1 + \ldots x_n y_n \pmod 2$.

Prove that:
(a) $\mathscr{C}^\perp$ is an $[n, n - k]$-code;
(b) if $\mathscr{C}$ has generator matrix $G = [I_k, A]$, then $\mathscr{C}^\perp$ has generator matrix $[A^T, I_{n-k}]$.

17. How would you construct a binary $[30, 11, 6]$-code? how many codewords does it have and what are its error-correcting capabilities?

18. Prove that, if $g(x) = g_0 + g_1 x + \ldots + g_k x^k \ne 0$ is the generator polynomial of a cyclic code, then $g_0 \ne 0$.

19. Show that the Reed–Muller code $\mathscr{C}(1, m)$ is the same code as the

Hadamard code obtained by forming the m-fold Kronecker product of

$$H_2 = \begin{bmatrix} 1 & 1 \\ -1 & 1 \end{bmatrix}$$

with itself.

20. Let $\mathscr{C}_1$ and $\mathscr{C}_2$ be respectively $[n_1, k_1, d_1]$ and $[n_2, k_2, d_2]$ linear binary codes with generator matrices G_1 and G_2 respectively. Show that the Kronecker product $G_1 \otimes G_2$ is the generator matrix of a binary linear code $\mathscr{C}_1 \otimes \mathscr{C}_2$ which has parameters $[n_1 n_2, k_1 k_2, d_1 d_2]$. ($\mathscr{C}_1 \otimes \mathscr{C}_2$ is called the *direct product* of $\mathscr{C}_1$ and $\mathscr{C}_2$.)

21. Show that, if

(a) $2^k \sum_{i=0}^{d-2} \binom{n-1}{i} < 2^n$,

then there exists a binary linear $[n, k]$-code with minimum distance at least d.

Hence deduce that

(b) $A_2(n, d) \geq 2^k$,

where k is the largest integer satisfying the inequality (a).

Hint: Construct an $(n - k) \times n$ matrix H with the property that no $d - 1$ of its columns are independent. By Problem 7, this gives the required code.

To achieve this, pick successively vectors from V_n which are to be columns of H. Keep choosing successive columns so that each new column is not a linear combination of any $d - 2$ or fewer previous columns.

Note: This is also called the Gilbert–Varshamov bound and is an improvement on the elementary bound of Section 2. For example this bound gives $A_2(10, 3) \geq 64$, as against the lower bound of 19 derived from (2.2).

22. Let $\mathscr{C}$ be a binary (n, M, d) code with $n < 2d$. Show that

$$M(M - 1)d \leq \sum\sum d(c_i, c_j) \leq \tfrac{1}{2} nM^2,$$

where the summation is over all codewords c_i and c_j of $\mathscr{C}$.

23. Use the previous question to show that: (a) if d is even and $2d > n$, then

$$A_2(n, d) \leq 2 \left\lfloor \frac{d}{2d - n} \right\rfloor, \qquad A_2(2d, d) \leq 4d.$$

(b) is d is odd and $2d + 1 > n$, then

$$A_2(n, d) \leq \left\lfloor \frac{d + 1}{2d + 1 - n} \right\rfloor, \qquad A_2(2d + 1, d) \leq 4d + 4.$$

Note 1: These results are what are known as Plotkin's bound and follow easily from the previous question and Problems 4 and 7.

(Plotkin, 1960)

Note 2: A result of Levenshtein (1961) says that, provided enough Hadamard matrices exist, equality holds in the Plotkin bounds above; the proof is constructive and based on the Hadamard codes of Section 8.

24. The Reed–Muller code $\mathscr{C}(1, 5)$ is used to transmit across a binary symmetric channel of bit error probability $p = 0.05$. What is the probability that any given block is incorrectly decoded? What is the rate of the code?

Note: This is the (32, 64, 16)-code which was used by NASA on its spacecraft from 1969 to 1976.

5

General sources

5.1 The entropy of a general source

The model of a zero-memory source is clearly unrealistic for a natural language such as English. In this chapter, we will try to extend some of the concepts discussed earlier to a wider class of sources.

First recall the definition of a source $\mathscr{S}$. In its widest sense, this is just an object that emits symbols from a finite alphabet Σ according to some random mechanism. We will let $(X_1, X_2, \ldots)$ denote the random elements of Σ emitted by $\mathscr{S}$, where X_i denotes the ith symbol emitted.

Clearly the n-stage entropy $H(X_1, \ldots, X_n)$ is well defined, and we say $\mathscr{S}$ *has entropy* $H(\mathscr{S}) = H$ if

$$\lim_{n \to \infty} \frac{H(X_1, \ldots, X_n)}{n}$$

exists and equals H. Obviously, if $\mathscr{S}$ is just a zero-memory source, then this definition reduces to the usual one since, by the zero memory property, we have

$$H(X_1, \ldots, X_n) = H(X_1) + \ldots + H(X_n).$$

At the other extreme, consider the 'highly dependent' source of the following example.

Example $\mathscr{S}$ emits either a string of zeros or a string of ones, each with probability $\frac{1}{2}$. Then, for this source, $H(X_1, \ldots, X_n) = 1$ so that $H(\mathscr{S}) = 0$. □

An alternative possible definition for the entropy of a source is to take the limit of $H(X_n \mid X_1, \ldots, X_{n-1})$ as n tends to infinity. Luckily we have the following theorem.

Theorem 1 *If the source $\mathscr{S}$ is such that*

$$\lim_{n \to \infty} H(X_n \mid X_1, \ldots, X_{n-1})$$

exists, then

$$\lim_{n\to\infty} \frac{H(X_1, \ldots, X_n)}{n}$$

exists and the two limits are equal.

Note The converse to this theorem does not hold: see Exercise 3 below.

The proof of the theorem depends on the following elementary lemma, whose proof we leave to the reader.

Lemma (Arithmetic mean lemma) *If $(a_n : n \geq 1)$ is a sequence of real numbers such that $a_n \to A$ as $n \to \infty$ then*

$$\lim_{n\to\infty} \frac{a_1 + \ldots a_n}{n} = A.$$

Using this lemma, it is not difficult to prove Theorem 1.

Proof of Theorem 1 Write

$$u_n = H(X_n \mid X_1, \ldots, X_{n-1}), \qquad h_n = H(X_1, \ldots, X_n),$$

so that, using $H(U, V) = H(U) + H(V \mid U)$, we get

$$h_n = h_{n-1} + u_n.$$

Hence $h_n = u_1 + \ldots + u_n$, and so

$$\lim_{n\to\infty} \frac{h_n}{n} = \lim_{n\to\infty} \frac{u_1 + \ldots + u_n}{n}$$

whenever the right hand exists. By the preceding lemma, the right-hand limit exists whenever $\lim_{n\to\infty} u_n$ exists, and all the limits are equal. $\square$

Exercises 5.1 1. A source $\mathcal{S}$ is such that, for each n, the probability distribution of X_n depends on X_{n-1} in the following way:

$$P(X_n \mid X_{n-1}, X_{n-2}, \ldots, X_1) = P(X_n \mid X_{n-1}),$$

$$P(X_n = 0 \mid X_{n-1} = 0) = P(X_n = 1 \mid X_{n-1} = 1) = p,$$

$$P(X_n = 1 \mid X_{n-1} = 0) = P(X_n = 0 \mid X_{n-1} = 1) = q,$$

where p and q satisfy $0 \leq p, q \leq 1$ and $p + q = 1$. Show that $H(\mathcal{S})$ exists and find it.

2. A source $\mathcal{S}$ behaves as follows: it emits, each with probability $\frac{1}{2}$, either an infinite string of zeros or a purely random string of zeros and ones. Does $\mathcal{S}$ have an entropy?

3. Consider a source $\mathscr{S}$ whose output $(X_n : n = 1, 2, \ldots)$ is as follows: $X_{2k} = 1$ for all k, and $(X_1, X_3, X_5, \ldots)$ is a purely random source. Show that $H(\mathscr{S})$ exists but that $\lim_{n \to \infty} H(X_n \mid X_1, \ldots, X_{n-1})$ does not.

5.2 Stationary sources

In order for us to be able to have some control over the amount of dependence between successive elements in the output of a source that is not memoryless, we need to place some restrictions on it. A fairly natural restriction is that it be *stationary*. This means that, for any positive integers n and h and for any sequence $(s_1, \ldots, s_n)$ of symbols from the output alphabet Σ, and for any non-negative indices $i_1, \ldots, i_n$, we must have

(1)
$$P(X_{i_1} = s_1, \ldots, X_{i_n} = s_n) = P(X_{i_1+h} = s_1, \ldots, X_{i_n+h} = s_n).$$

Mathematically, this is a fairly strong restriction to put on a source. It essentially means that the probability that a given sequence is emitted is independent of time. In particular, the probability that a particular symbol 's' is emitted is a constant over time; similarly, the probability that a given digram '$s_1 s_2$' is emitted is independent of time and so on. As a model of a natural language, such as English, it is plausible. Take a novel and regard it as a source; then, provided we don't start at the beginning of a sentence where perhaps the letter 'T' is too probable, the stationary model seems to be a reasonable approximation. We return to this in Chapter 6.

As is to be expected, we can prove the following result.

Theorem 1 *Any stationary source has an entropy.*

Our proof of this depends upon the following property of subadditive functions. It is a remarkably useful lemma which deserves to be more widely known.

A function $f : \mathbb{Z}^+ \to \mathbb{R}$ is said to be *subadditive* if, for all $x, y \in \mathbb{Z}^+$,

(2)
$$f(x + y) \leq f(x) + f(y).$$

FUNDAMENTAL LEMMA ON SUBADDITIVE FUNCTIONS

If f is a subadditive function on $\mathbb{Z}^+$, then

(3)
$$\lim_{n \to \infty} \frac{f(n)}{n} = \inf_{n \geq 1} \frac{f(n)}{n}$$

exists and is finite.

The proof is not difficult; the interested reader is referred to Hille and Phillips (1957), where a more general version is given. An example of one of the diverse applications of this result is shown in Exercise 1 below.

Using this fundamental lemma, we get a quick proof of Theorem 1.

Proof Let $X_1, X_2, \ldots$ denote the output of the stationary source and let

$$g(n) = \mathrm{H}(X_1, \ldots, X_n).$$

Then

$$g(m + n) = \mathrm{H}(X_1, \ldots, X_{m+n})$$
$$\leq \mathrm{H}(X_1, \ldots, X_m) + \mathrm{H}(X_{m+1}, \ldots, X_{m+n}).$$

But, by stationarity,

$$\mathrm{H}(X_{m+1}, \ldots, X_{m+n}) = \mathrm{H}(X_1, X_2, \ldots, X_n)$$
$$= g(n).$$

Thus,

$$g(m + n) \leq g(m) + g(n)$$

and, since g is non-negative, the result follows. $\square$

Corollary *The entropy of a stationary source $\mathscr{S}$ is bounded above by*

$$\mathrm{H}(\mathscr{S}) \leq n^{-1}\mathrm{H}(X_1, \ldots, X_n)$$

for any positive integer n.

Proof This is immediate from the property of subadditive functions that says (in terms of the preceding proof)

$$\lim_{n \to \infty} n^{-1}g(n) = \inf_n n^{-1}g(n).$$

$\square$

Theorem 2 *If $\mathscr{S}$ is a stationary source then $\lim_{n \to \infty} \mathrm{H}(X_n \mid X_1, \ldots, X_{n-1})$ exists and equals the entropy of $\mathscr{S}$.*

Thus, for stationary sources, we have an extension of Theorem 1 of Section 5.1.

Proof $\mathrm{H}(X_n \mid X_1, \ldots, X_{n-1}) \leq \mathrm{H}(X_n \mid X_2, X_3, \ldots, X_{n-1})$ by the basic property of entropy. But, by stationarity,

$$\mathrm{H}(X_n \mid X_2, X_3, \ldots, X_{n-1}) = \mathrm{H}(X_{n-1} \mid X_1, \ldots, X_{n-2}).$$

Hence

$$u_n = H(X_n \mid X_1, \ldots, X_{n-1}) \le H(X_{n-1} \mid X_1, \ldots, X_{n-2})$$

$$= u_{n-1}.$$

Thus u_n is monotone decreasing, and hence converges, as $n \to \infty$. The result now follows from Theorem 5.1.1. $\square$

Exercises 1. Consider the infinite square lattice consisting of all integer-coordinated
5.2   points of the plane and with nearest neighbours in the direction of the
              coordinate axes joined by an edge. A *self-avoiding walk* is a sequence of
              n edges starting from the origin, each pair of consecutive edges having a
              common point, and at no stage revisiting a point already visited. If $f(n)$
              denotes the number of self-avoiding walks having n edges, then clearly
              $f(1) = 4$, $f(2) = 12$, and so on. Prove that

$$f(m + n) \le f(m)f(n),$$

and hence deduce that

$$\lim_{n \to \infty} [f(n)]^{1/n} = \inf_{n \ge 1}[f(n)]^{1/n} = \theta$$

exists.

Prove further that $2 \le \theta \le 3$.

Note: An exact evaluation of θ is an unsolved problem of long standing.

2. Each of the Exercises 1, 2, and 3 of the previous section describe a
   source. Which, if any, of these sources is stationary?

5.3   Typical messages of a memoryless source

Suppose that we have a source which is a mathematical model of
English. Any such source is probably going to have a positive
probability of emitting a string of 20 successive A's. However, this
probability will be very low and our 'typical' output, hopefully, will
approximate to English with roughly speaking between 10% and 20%
of E's, frequent spaces, few Z's and so on. In this section we follow
Shannon's idea of dividing source outputs into those that are 'typical'
and those that are not.

Although the main result will hold for sources that are more
general, we have restricted attention here to the memoryless case in
order to make both the statement and proof more readily
understood.

Theorem Let $\mathscr{S}$ be a memoryless source with entropy H. Then, given any $\varepsilon > 0$,
1   the set $\Sigma^{(N)}$ of sequences of length N can be divided into two classes:

(1)   *A set Π_N such that, if X_N denotes the random output of length N of $\mathcal{S}$, then*

$$P(X_N \in \Pi_N) < \varepsilon.$$

(2)   *The remainder, $\Sigma^{(N)} \backslash \Pi_N$, all of whose members σ_N have 'high probability' satisfying the inequality*

$$2^{-NH-AN^{\frac{1}{2}}} \le P(X_N = \sigma_N) \le 2^{-NH+AN^{\frac{1}{2}}},$$

where A is a positive constant.

In other words, $\Sigma^{(N)}$ consists of a set of *low probability* or *atypical* sequences (namely Π_N) and a disjoint set of *high probability* or *typical* sequences each of which has a probability of occurrence approximately $\frac{1}{2}^{NH}$.

Proof   Let the source alphabet $\Sigma = \{s_1, \ldots, s_n\}$. If $x_N = (x_1, \ldots, x_N)$ is any sequence of symbols from Σ, we say that x_N is *typical* if

(3)
$$\left| \frac{f_i(x_N) - Np_i}{(Np_iq_i)^{\frac{1}{2}}} \right| \le W \quad (1 \le i \le n),$$

where p_i is the probability of emitting symbol s_i, with $q_i = 1 - p_i$, and where $f_i(x_N)$ is the frequency (number) of occurrences of symbol s_i in x_N; here, W is (for the moment) some arbitrary positive quantity.

Let $\Pi_N(W)$ be the subset of $\Sigma^{(N)}$ consisting of atypical sequences, that is, sequences x_N violating (3). Now consider a random output $X_N = (X_1, \ldots, X_N)$ from $\mathcal{S}$. Using the basic rule $P(A \cup B) \le P(A) + P(B)$, we get

(4)
$$P(X_N \in \Pi_N(W)) = P\left(\bigcup_{i=1}^{n} \left\{ \left| \frac{f_i(X_N) - Np_i}{\sqrt{Np_iq_i}} \right| > W \right\} \right)$$

$$\le \sum_{i=1}^{n} P\left\{ \left| \frac{f_i(X_N) - Np_i}{\sqrt{Np_iq_i}} \right| > W \right\}$$

$$\le \sum_{i=1}^{n} \frac{1}{W^2} E\left(\frac{|f_i(X_N) - Np_i|^2}{Np_iq_i} \right)$$

by Chebyshev's inequality (see §3.5).

But consider the quantity $f_i(X_N)$. This is just a random variable counting the number of occurrences of s_i in N symbols and so

(5) $f_i(X_N)$ is binomially distributed with parameters N and p_i.

Hence
$$E\, |f_i(X_N) - Np_i|^2 = Np_iq_i,$$

so that substituting back in (4) gives

$$P(X_N \in \Pi_N(W)) \le n/W^2.$$

But n is fixed, and hence if initially we had chosen W so that $n/W^2 = \varepsilon$ we have a class Π_N such that

$$P(X_N \in \Pi_N) \le \varepsilon$$

as required by (1).

To prove the second assertion, suppose that σ_N is a typical sequence of length N, that is, satisfying (3). Then, for all i $(1 \le i \le n)$, we have

(6)
$$Np_i - W(Np_iq_i)^{\frac{1}{2}} \le f_i(\sigma_N) \le Np_i + W(Np_iq_i)^{\frac{1}{2}}.$$

But

$$P(X_N = \sigma_N) = p_1^{f_1(\sigma_N)} p_2^{f_2(\sigma_N)} \ldots p_n^{f_n(\sigma_N)}.$$

Hence substituting from (6) and writing $-W \sum (p_iq_i)^{\frac{1}{2}} \log p_i = A$ gives

(7)
$$2^{-NH-AN^{\frac{1}{2}}} \le P(X_N = \sigma_N) \le 2^{-NH+AN^{\frac{1}{2}}},$$

as required.   □

An immediate consequence of the theorem is the following.

Corollary   *The number of typical sequences of length N emitted by a memoryless source of entropy H is*

(8)
$$2^{NH+o(N)} \quad \text{as } N \to \infty.$$

Proof   Since the total number of N-sequences is 2^N, the number $t(N)$ of typical sequences satisfies, by taking their total probability and using (7), the inequality

$$t(N)/2^{NH+AN^{\frac{1}{2}}} \le 1.$$

Thus

$$t(N) \le 2^{NH+AN^{\frac{1}{2}}},$$

where A is the constant referred to in the proof above. Similarly, by bounding each probability below, we get the reverse inequality

$$t(N) \ge (1 - \varepsilon)2^{NH-AN^{\frac{1}{2}}},$$

and this proves the corollary.   □

Exercises   1. A memoryless source emits only vowels, each with the following probabilities:

$$P(A) = 0.2, \qquad P(E) = 0.3, \qquad P(I) = P(O) = 0.2, \qquad P(U) = 0.1.$$

Estimate the number of typical outputs of length n.

2. A memoryless source over the 26-letter alphabet has a vocabulary of about 10^n n-letter sequences, for sufficiently large n. Estimate the entropy of the source.

5.4 Typical messages of general sources—ergodicity

Suppose we try to generalize the idea of a 'typical sequence' of source symbols to more general sources. Our aim would be to find a way of dividing the set $\Sigma^{(N)}$ of all possible N-sequences from Σ into two disjoint sets: the typical group $T(\Sigma^{(N)})$ and the low-probability group consisting of the rest. Moreover, within this typical set, each N sequence should have roughly the same probability 2^{-NH}, where H is the source entropy.

First note that we cannot achieve anything approaching this for a general stationary source, as the following example shows.

Example 1 Suppose a source $\mathscr{S}$ emits $000\ldots$ with probability p or it emits a random binary string with probability $1-p$. This is a stationary source and, if $X_1, \ldots, X_N$ denotes the output, then $H(X_1, \ldots, X_N) = N(1-p)$, so that

$$H(\mathscr{S}) = (1-p).$$

However, there is no way of dividing the possible N-sequences into typical and atypical groups in which the typical sequences have roughly the same probability, because $000\ldots$ is much more likely than any other sequence. $\qquad\square$

In order to be able to do this, we need the concept of ergodicity. This is a difficult concept, but we will attempt to give the reader an informal idea of what it means. We say that the source $\mathscr{S}$ with entropy H has the *asymptotic equipartition property* (AEP) if for each value of N it is possible to partition the collection $\Sigma^{(N)}$ of N-sequences of its source alphabet Σ into a *typical set* $T(\Sigma^{(N)})$ and a *low-probability set* $\Pi(\Sigma^{(N)})$ such that each typical sequence must have probability approximately equal to 2^{-NH}, while the total probability of the low-probability group is arbitrarily small for sufficiently large N. As Example 1 shows, not all stationary sources have the AEP property.

We will define a source $\mathcal{S}$ over Σ to be *ergodic* if it is stationary and if it also has the following property (1) below. For any sequence $s = (s_1, \ldots, s_a)$ of symbols of Σ and any output $X = (X_1, X_2, \ldots)$ of the source, define the *frequency* $f_N(s, X)$ to be the number of times s occurs in the first N terms of the sequence X. For example, if

$$s = 011 \quad \text{and} \quad X = 0010110011011001,$$

then $f_{16}(s, X) = 3$.

Thus $\mathcal{S}$ is *ergodic* if it is stationary and if, for any finite sequence $s = (s_1, \ldots, s_a)$ of symbols from Σ, we have

(1)
$$\lim_{N \to \infty} N^{-1} f_N(s, X) = P(X_1 = s_1, \ldots, X_a = s_a).$$

Note 1 We emphasize that, in the definition (1), the left-hand side is a random variable and the right-hand side is not. What we mean by (1) is that

$$P\left(\lim_{N \to \infty} \frac{f_N(s, X)}{N} = P(X_1 = s_1, \ldots, X_a = s_a) \right) = 1.$$

To those familiar with convergence concepts in probability, this will be recognizable as demanding that $f_N(s, X)/N$ converges with probability 1 to the required constant.

Note 2 In the definition (1), because $\mathcal{S}$ is stationary, the right-hand probability could be replaced by $P(X_{b+1} = s_1, \ldots, X_{b+a} = s_a)$ for any integer $b > 0$.

For ergodic sources, we have the following fundamental theorem known as the Shannon–McMillan Theorem.

Theorem 1 *Ergodic sources have the asymptotic equipartition property.*

More precisely this can be stated as follows.

Theorem 1′ *Let $\mathcal{S}$ be an ergodic source with entropy H. Then, for any $\varepsilon > 0$, there exists a positive integer $N_0(\varepsilon)$ such that, if $N > N_0(\varepsilon)$, the set $\Sigma^{(N)}$ of possible N-sequences of the source alphabet decomposes into two sets Π and T satisfying*

(a) $$P(X_N \in \Pi) < \varepsilon,$$

(b) $$2^{-N(H+\varepsilon)} < P(X_N = \sigma) < 2^{-N(H-\varepsilon)}$$

for any N-sequence $\sigma \in T$.

A rigorous proof is tricky; we refer to Billingsley (1965).

It is clearly nontrivial to prove that a source is ergodic. For example the reader may readily believe the statement

(2)   Every memoryless source is ergodic.

Proving this result is essentially proving a law of large numbers in probability theory. It is usually easier to recognise non-ergodic sources. We will however meet a large class of ergodic sources in the next section.

Exercises 1. With probability $\frac{1}{3}$, a source $\mathcal{S}$ emits a random string of zeros and ones;
5.4   with probability $\frac{2}{3}$, it emits a random string of ones and twos. Show that this source is not ergodic.

5.5 Markov sources

As we will see in the next chapter, Markov sources are probably the most realistic simple models of natural languages.

A source $\mathcal{S}$ with alphabet $\Sigma = \{s_1, \ldots, s_m\}$ is said to be a *Markov source* if, letting $X = X_1, X_2, \ldots$ represent the output of the source, then, for each $n \geq 1$ and any collection of source symbols s_j, s_i, $s_h, \ldots, s_\alpha$, the following property holds:

(1)
$$P(X_{n+1} = s_j \mid X_n = s_i, X_{n-1} = s_h, \ldots, X_1 = s_\alpha)$$
$$= P(X_{n+1} = s_j \mid X_n = s_i).$$

We then denote this probability $P(X_{n+1} = s_j \mid X_n = s_i)$ by p_{ij}. In other words, the probability distribution of X_{n+1} given the past history depends only on the most recent output X_n.

Clearly, in order for this to be well defined, we must have $p_{ij} \geq 0$ and

(2)
$$\sum_j p_{ij} = 1 \quad (1 \leq i \leq m).$$

the matrix P whose (i, j)th entry is p_{ij} is called the *transition matrix* of the source.

In order that the source be completely specified, we also need to give the *initial probabilities*

(3)
$$\pi_i = P(X_1 = s_i) \quad (1 \leq i \leq m).$$

Readers familiar with stochastic processes will recognize the above definitions as thinly disguised finite-state Markov chains.

Suppose now we wish to work out the entropy of this Markov

source. Because of its Markovian nature, it is natural to consider

$$H(X_{n+1} \mid X_n) = \sum_{j=1}^{m} H(X_{n+1} \mid X_n = s_j)P(X_n = s_j).$$

The entropies $H(X_{n+1} \mid X_n = s_j)$ are easy to work out: they are given by

(4)
$$H(X_{n+1} \mid X_n = s_j) = -\sum_{k=1}^{m} p_{jk} \log p_{jk} = H_j.$$

However, the terms $P(X_n = s_j)$ depend on the initial distribution $\{\pi_i\}$.

Accordingly, we consider the following recurrence relation. Define the *absolute probability* $a_j(n)$ by

$$a_j(n) = P(X_n = s_j).$$

Then, using basic conditional probability and because of the Markov property (1), we have

(5)
$$a_j(n + 1) = \sum_i P(X_{n+1} = s_j \mid X_n = s_i)P(X_n = s_i)$$

$$= \sum_i p_{ij}a_i(n),$$

$$a_j(1) = \pi_j.$$

Whenever

$$\lim_{n \to \infty} a_j(n) = a_j$$

exists for all j with $1 \le j \le m$, we say that $\boldsymbol{a} = (a_1, \ldots, a_m)$ is the *steady-state distribution* of the source $\mathscr{S}$ and, in this case, if we write

$$H(X_{n+1} \mid X_n) = \sum_{j=1}^{m} H_j a_j(n)$$

and then take the limit in this as $n \to \infty$, we get that the limit

$$\lim_{n \to \infty} H(X_{n+1} \mid X_n) = \sum_{j=1}^{m} a_j H_j$$

exists, and the right-hand side is the entropy of the Markov source $\mathscr{S}$. In other words we have proved the next result.

Theorem 1 *If a Markov source $\mathscr{S}$ has a steady-state distribution $(a_k : 1 \le k \le m)$, then its entropy H is given by*

$$H = -\sum_i a_i \sum_j p_{ij} \log p_{ij}.$$

In order to use Theorem 1 effectively, we need to be able to find the steady-state distribution of the source whenever it exists. This we do as follows. Consider the recurrence equations (5). Since the sum is finite, then, by taking the limit as $n \to \infty$, the steady state distribution $a = (a_1, \ldots, a_m)$ must satisfy the set of homogeneous equations

(6)
$$a_j = \sum_{i=1}^{m} p_{ij} a_i \quad (1 \le j \le m).$$

We also know that, since a is a probability distribution,

(7)
$$\sum_{i=1}^{m} a_i = 1.$$

For a general Markov source these equations may not have a unique solution. However we will show that provided $\mathscr{S}$ satisfies a very natural condition they can be used to calculate the entropy.

The matrix P is *irreducible* if for each i, j, there exists n such that $(P^n)_{ij} > 0$. In terms of the source $\mathscr{S}$, this can be interpreted as follows. It is easy to see that for any positive integers n, k,

$$(P^n)_{ij} = P(X_{n+k} = s_j \mid X_k = s_i).$$

Hence P is irreducible if the source $\mathscr{S}$ is such that for each pair of symbols s_i and s_j there is a non-zero probability that at some time following the appearance of s_i the source will emit s_j.

In any situation where a Markov source is used to model a natural language such as English this is surely true.

When the above condition is satisfied we describe the source $\mathscr{S}$ as irreducible and for such sources we have the following basic result from the theory of Markov chains, see Grimmett and Stirzaker (1982).

(8)
If the Markov source $\mathscr{S}$ is irreducible then the equations (6) and (7) have a unique non-negative solution $v = (v_1, \ldots, v_m)$, which is called the *stationary distribution* of the source.

Clearly when there is a steady-state distribution it must be the stationary distribution. More generally we have that for any irreducible Markov source,

(9)
$$\lim_{n \to \infty} n^{-1} \sum_{k=1}^{n} a_i(k) = v_i \quad (1 \le i \le m).$$

If we now use the Markov property of $\mathscr{S}$ to write

(10)
$$H(X_1, \ldots, X_n) = H(X_1) + \sum_{k=2}^{n} H(X_k \mid X_{k-1})$$

we can combine (4), (9), and (10) to obtain

$$\lim_{n \to \infty} n^{-1} H(X_1, \ldots, X_n) = \sum_{i=1}^{m} v_i H_i.$$

In other words we have proved the following statement.

(11)   An irreducible Markov source has an entropy H given by

$$H = \sum_{i=1}^{m} v_i H_i$$

where $v = (v_1, \ldots, v_m)$ is the stationary distribution and

$$H_i = -\sum_j p_{ij} \log p_{ij}.$$

Example Suppose a Markov source has alphabet $\Sigma = \{0, 1\}$ and has a transition matrix

$$P = \begin{bmatrix} \frac{1}{4} & \frac{3}{4} \\ \frac{3}{4} & \frac{1}{4} \end{bmatrix},$$

so that each symbol has only probability $\frac{1}{4}$ of being succeeded by itself. The individual symbol entropies are equal and given by

$$H_0 = H_1 = \tfrac{1}{4} \log 4 + \tfrac{3}{4} \log \tfrac{4}{3}$$
$$= \log 4 - \tfrac{3}{4} \log 3.$$

The stationary distribution is got by solving

$$a_1 = \tfrac{1}{4} a_1 + \tfrac{3}{4} a_2, \qquad a_2 = \tfrac{3}{4} a_1 + \tfrac{1}{4} a_2, \qquad a_1 + a_2 = 1,$$

giving $a_1 = a_2 = \tfrac{1}{2}$, so that the source entropy is also $\log 4 - \tfrac{3}{4} \log 3$. $\square$

Now, despite all these nice properties, an irreducible Markov source will not be stationary *unless* the unique stationary distribution is used as the initial distribution. However, when this is the case, the source is not only stationary but ergodic.

This is really a fundamental theorem in the theory of finite Markov chains which goes back to A. A. Markov. However, because of its importance, we restate it here.

Theorem 2 *If $\mathscr{S}$ is an irreducible Markov source, and if the unique stationary distribution is taken as the initial distribution, then $\mathscr{S}$ is an ergodic source which therefore has the AEP property.*

Note In practical situations, it is unlikely that the initial distribution will be exactly the stationary distribution. However, if we let the source run for a long time before starting the time clock, then it is reasonable to expect the initial distribution to be close to the steady state whenever the latter exists.

Exercises 5.5

1. Find the entropy of the Markov source whose transition matrix is given by

$$\begin{bmatrix} \frac{1}{3} & \frac{1}{3} & \frac{1}{3} \\ \frac{1}{2} & 0 & \frac{1}{2} \\ 0 & \frac{1}{2} & \frac{1}{2} \end{bmatrix}.$$

2. Which of the Markov sources having transition matrices as shown are irreducible?

(a) $\begin{bmatrix} \frac{1}{3} & \frac{1}{3} & \frac{1}{3} \\ \frac{1}{3} & \frac{1}{3} & \frac{1}{3} \\ \frac{1}{3} & \frac{1}{3} & \frac{1}{3} \end{bmatrix}$, (b) $\begin{bmatrix} 1 & 0 & 0 \\ \frac{1}{3} & \frac{1}{3} & \frac{1}{3} \\ 0 & \frac{1}{2} & \frac{1}{2} \end{bmatrix}$,

(c) $\begin{bmatrix} \frac{1}{2} & \frac{1}{2} & 0 & 0 \\ \frac{1}{2} & \frac{1}{2} & 0 & 0 \\ 0 & 0 & \frac{1}{2} & \frac{1}{2} \\ 0 & 0 & \frac{1}{4} & \frac{3}{4} \end{bmatrix}$, (d) $\begin{bmatrix} 0 & \frac{1}{2} & 0 & \frac{1}{2} \\ \frac{1}{2} & 0 & \frac{1}{2} & 0 \\ 0 & \frac{1}{2} & 0 & \frac{1}{2} \\ \frac{1}{2} & 0 & \frac{1}{2} & 0 \end{bmatrix}.$

3. Which initial distribution would make the source in Exercise 1 an ergodic source?

5.6 The coding theorems for ergodic sources

We conclude this chapter with a brief sketch of the way in which Shannon's two major theorems can be extended from the case of a memoryless source to that of any source which has the AEP property, and in particular to all ergodic sources.

Hence, suppose $\mathscr{S}$ is a source with the AEP property and having entropy H.

Then, for sufficiently large N, we know that the sequences of length N can be divided into the high-probability group of typical sequences and the low-probability group of atypical sequences.

THE NOISELESS CODING THEOREM

Suppose we wish to encode the output of $\mathscr{S}$ into an alphabet Σ of size D.

For a given $\varepsilon > 0$, let N be large enough for the number of typical strings T_N of length N to satisfy

$$T_N \leq 2^{N(H+\varepsilon)}.$$

Encode each of these by a distinct string of length r of symbols from the source alphabet of size D. Since there are D^r such strings, this can be done when

$$D^r > T_N,$$

that is, when

$$r \log D \geq N(H + \varepsilon).$$

The other, atypical, sequences from the source $\mathcal{S}$ we encode by first prefixing a fixed string σ_0 of length r that was not used in the encoding of the typical sequences, and then encoding them by a string of length N. This gives an encoding of average length not greater than

$$\frac{N(H + \varepsilon)}{\log D} + \delta N,$$

where δ is the probability that $\mathcal{S}$ outputs an atypical sequence.

Thus, if l_N denotes the average length of an encoding of an N sequence of symbols from the source, then

$$\frac{l_N}{N} \leq \frac{H}{\log D} + \delta'$$

and we have found a 'compact encoding' of $\mathcal{S}$. $\square$

LINKING AN ERGODIC SOURCE TO THE BINARY SYMMETRIC CHANNEL

Suppose again we have an ergodic source $\mathcal{S}$ of entropy H which we propose to link to a binary symmetric channel of capacity C. Provided that $H < C$, we can find R such that

$$H < R < C$$

and, by the noisy coding theorem for binary symmetric channels (Theorem 2), we know that there exists a sequence of codes $(\mathcal{C}_n : 1 \leq n < \infty)$ such that $\mathcal{C}_n$ has 2^{Rn} codewords of length n and error probability that tends to zero as $n \to \infty$.

Given this sequence of codes, we take N large enough so that the number of typical strings emitted by $\mathcal{S}$ is $\approx 2^{NH}$. Then, since $H < R$, we can encode them by the $\lfloor 2^{NR} \rfloor$ codewords of $\mathscr{C}_N$ and encode the remainder of the atypical N-strings of $\mathscr{S}$ arbitrarily, thus achieving an encoding with low probability of error. □

While the above discussion does not pretend to be a proof in the technical sense of either of the above theorems (for full details see Khinchin, 1957, or Feinstein, 1958), it may give the reader some flavour of the underlying idea. In both cases the basic principle is the same, namely the following.

Encoding principle

Encode the high-probability typical sequences carefully.
Encode the low-probability atypical sequences arbitrarily.

PROBLEMS 5

1. A memoryless source with alphabet $\Sigma = \{0, 1\}$ emits 0 with probability p, and 1 with probability $1 - p$. If the number of typical strings of length n is approximately $(\frac{3}{2})^n$ for large n, what is your estimate for p?
2. A memoryless source $\mathscr{S}$ emits symbols in bursts of four. The probability distribution of an output is $p(x)$, where

$$p(0000) = \tfrac{1}{2}, \qquad p(1100) = \tfrac{1}{4}, \qquad p(1110) = \tfrac{1}{4}.$$

What is the entropy of $\mathscr{S}$?
   If $\mathscr{S}$ is regarded as a source $\mathscr{S}_1$ over the binary alphabet $\{0, 1\}$, then is $\mathscr{S}_1$ stationary or ergodic? How is the entropy of $\mathscr{S}_1$ related to that of $\mathscr{S}$?
3. A Markov source has transition matrix

$$\begin{bmatrix} \frac{1}{4} & \frac{1}{4} & \frac{1}{2} & 0 \\ \frac{1}{3} & 0 & \frac{1}{3} & \frac{1}{3} \\ 0 & \frac{1}{2} & \frac{1}{2} & 0 \\ \frac{1}{4} & \frac{1}{4} & \frac{1}{4} & \frac{1}{4} \end{bmatrix}.$$

Show that the source is ergodic if the initial distribution is chosen correctly and find its entropy.
4. Show that the Markov source with transition matrix

$$P = \begin{bmatrix} \frac{1}{2} & \frac{1}{2} & 0 & 0 \\ \frac{1}{2} & \frac{1}{2} & 0 & 0 \\ 0 & 0 & \frac{1}{4} & \frac{3}{4} \\ 0 & 0 & \frac{3}{4} & \frac{1}{4} \end{bmatrix}$$

is not irreducible. What is the value of its entropy?
5. Let $\mathscr{S}$ be a source with alphabet $\Sigma = \{0, 1, 2\}$ and suppose that $\mathscr{S}$ operates as follows. With probability p, the source $\mathscr{S}$ emits a string of

zeros; with probability $1-p$, it emits a random string of 1's and 2's, chosen independently and with equal probability. Show that:
(a) $H(\mathcal{S})$ exists and equals $1-p$,
(b) there exists a uniquely decipherable encoding $\mathscr{C}_n$ of $\Sigma^{(n)} \rightarrow \{0, 1\}$ such that the average length $l(\mathscr{C}_n)$ of the encoding satisfies

$$n^{-1}l(\mathscr{C}_n) \rightarrow 1 - p \quad \text{as } n \rightarrow \infty.$$

6. If $\mathcal{S}$ is an ergodic Markov source with alphabet $\Sigma = \{s_1, \ldots, s_n\}$ and with stationary distribution $(a_1, \ldots, a_n)$, denote by $\mathcal{S}^*$ the memoryless source with alphabet Σ and probability a_i of emitting s_i $(1 \leq i \leq n)$. Show that

$$H(\mathcal{S}) \leq H(\mathcal{S}^*).$$

$(\mathcal{S}^*$ is sometimes called the *adjoint* of $\mathcal{S}$.)

7. Consider a source which at time 0 throws a fair die and, depending on the throw, chooses a parameter p to be $1/i$, where i is the throw of the die. It then proceeds to emit a binary sequence with each symbol independently being 0 with probability p, or 1 with probability $1-p$. Show that this is a stationary but not ergodic source. What is the value of its entropy?

6

The structure of natural languages

6.1 English as a mathematical source

In the previous chapter, we have defined various mathematical models of sources; we now investigate how closely these approximate the natural languages in everyday use.

Obviously, English has a very complicated structure, and an exact representation in mathematical terms is impracticable. However, we shall see that, depending on the type of English under discussion, we can get reasonable approximations to it by means of the concepts previously discussed.

We shall regard English (and the other languages discussed) as using the 27-letter alphabet consisting of the 26 Roman letters plus a space. A first, and very bad approximation to English is to take what is called the *zeroth-order approximation*. In this, all symbols are equally likely: each occurs with probability $\frac{1}{27}$ and the following display shows a typical sequence of symbols emitted by such a source.

(1)   DM QASCJDGFOZYNX ZSDZLXIKUD.

This approximation makes no use of the relative frequency of the symbols used in English. Using the estimates of these frequencies as listed in Appendix 2, we can produce a *first-order approximation* to English, a typical example of which is

(2)   OR L RW NILI E NNSBATEI.

Although this is more recognizable than the zeroth-order approximation, it still takes no account of the interdependencies between successive letters. We can accommodate this by setting up a *first-order Markov source*. In this, we could use the conditional probabilities based on frequencies of pairs of letters (that is, *digrams*):

$$p(i \mid j) = p(i, j)/p(j),$$

where $p(i, j)$ is the probability of the digram (i, j) and $p(i \mid j)$ is the conditional probability of an i given that the preceding letter is j. This is time-consuming to set up, and Shannon cleverly proposed a Monte Carlo method that has the same effect.

Pick a text, or texts, at random. Open at random and take the first letter as the first symbol X_1. Suppose it is B. Open another page at random, peruse until meeting the first B. Take our second symbol X_2 to be that letter in the text following B and so on. Using this method, we obtain the following first-order Markov approximation to English:

(3)   OUCTIE IN ARE AMYST TE TUSE SOBE CTUSE.

Shannon's method can be applied to obtain closer approximations, by selecting letters from a book according to the two preceding letters. Using his method, we obtained the following second-order Markov approximations to English:

(4)   HE AREAT BEIS HEDE THAT WISHBOUT SEED DAY OFTE,
       AND HE IS FOR THAT MINUMB LOOTS WILL AND GIRLS,
       A DOLL WILL IS FRIECE ABOARICE STRED SAYS,

(Source: *Runyon on Broadway*)

Using Shannon's method with Cicero's *de Senectute* as a source, we obtained the following very recognizable Markov approximations to Latin:

(5)   IENEC FES VIMONILLITUM M ST ER PEM ENIM PTAUL
                                              (first-order),

(6)   SENECTOR VCI QUAEMODOMIS SE NON
                 FRATURDIGNAVIT SINE VELIUS
                                              (second-order).

The same method when applied to French gave the second-order Markov approximation

(7)   MAITAIS DU VEILLECALCAMAIT DE LEU DIT,

and the third-order approximation

(8)   DU PARUT SE NE VIENNER PERDENT LA TET.

In theory, this method can be extended to obtain approximations of arbitrarily high order. However it is tiresome to do even a third-order approximation. (Try searching through a text for the next occurrence of "att" which is a not atypical trigram in English.)

However, the reader will, I hope, accept that the passage in Example 4 is a reasonable approximation to English. Approximations to other languages can be obtained in a similar way as shown in Examples 5–8.

An alternative approach suggested by Shannon was to model English not as a source of letters but as a source with the set of English *words* as the basic alphabet. Rather than use the table of frequencies of English words Shannon suggested using the 'open the book at random' method described above for letter frequencies. By this method he obtained the following approximations.

First-order word approximation

REPRESENTING AND SPEEDILY IS AN GOOD APT OR COME CAN DIFFERENT NATURAL HERE HE THE A IN CAME THE TO OF THE EXPERT GRAY COME TO FURNISHES THE LINE MESSAGE HAD BE THESE

Second-order word approximation

THE HEAD AND IN FRONTAL ATTACK ON AN ENGLISH WRITER THAT THE CHARACTER OF THIS POINT IS THEREFORE ANOTHER METHOD FOR THE LETTERS THAT THE TIME OF WHOEVER TOLD THE PROBLEM FOR AN UNEXPECTED.

I. J. Good (1969) proposed an alternative method of obtaining higher-order word approximations to a language. By asking colleagues chosen at random to supply the next word given only the previous three he obtained the following third-order word approximation to English:

"The best film on television tonight is there no-one here who had a little bit of fluff".

This method can obviously be extended to higher orders to obtain successively improving pieces of prose.

The conclusion of this section is therefore that it is not a ridiculous approximation to regard a natural language, such as English, as a limit of some succession of Markov sources as described above. Moreover, each of these sources will have associated with them a unique-steady state distribution, and since by their very nature they satisfy the conditions of Theorem 5.5.1, they will be ergodic sources. Thus, for the rest of this chapter we will assume that:

(9) English, and all other natural languages discussed, can be modelled by ergodic Markov sources.

Exercises 6.1

1. Prove that in the zeroth-order approximation to English, the average length of a word is 27 letters.
2. What is the average length of a word in the first-order approximation to English if the probability of a space is taken as 0.18?

6.2 The entropy of English

Because of our fundamental assumption about the ergodic nature of English and other natural languages, we can regard such languages as having an entropy. In this section, we try to give some estimates and interpretations of this entropy which, in the case of English, we shall denote by H_E.

By the Shannon–McMillan theorem, since we are regarding English as the output of an ergodic source, we can interpret H_E by using the approximate formula

(1)
$$2^{nH_E} \simeq T(n) \quad (n \text{ large})$$

where $T(n)$ denotes the number of typical (=meaningful) sequences of length n of English prose.

The relation (1) is not of immediate help in estimating H_E, since there is no way (known to the author) of estimating $T(n)$.

However, since there are 27^n possible sequences of n symbols from the 27-letter English alphabet, and since $\log 27 = 4.76$,

(2)
$$H_E \leq 4.76 \text{ bits per symbol.}$$

A better estimate of H_E is obtained from the first-order approximation, in which we take account of the differing probabilities that a letter will occur. For example, the most probable symbol is a space, with

$$P(\text{space}) = 0.18 \ldots, \qquad P(E) = 0.13 \ldots,$$

and so on.

Using the basic identity $H(X, Y) \leq H(X) + H(Y)$ gives an upper bound

$$H_E \leq H_E^1 = -\sum p_i \log p_i,$$

where p_i is the probability of occurrence of the ith symbol.

In the same way, we get a bound based on the frequency of digrams:

$$H_E \leq H_E^2 = -\tfrac{1}{2} \sum_i \sum_j p(i, j) \log p(i, j),$$

where $p(i, j)$ is the estimated probability of symbols (i, j), and we ignore pairings of zero probability such as Qq.

Table 1 lists Shannon's results using this method, extended to trigrams, and with both the 26- and 27-letter alphabets.

Table 1

n-gram entropies for n = 0, 1, 2, 3 (Shannon, 1951)

	26-letter alphabet	27-letter alphabet
H_E^0	4.70	4.76
H_E^1	4.14	4.03
H_E^2	3.56	3.32
H_E^3	3.30	3.10

An alternative approach of Shannon was based on estimating the conditional entropies $H(X_n | X_1, \ldots, X_{n-1})$.

This method was based on the reasonable assumption that an intelligent human subject understands his or her native language well enough to approximate the performance of an ideal predictor. Showing this subject the first $n-1$ letters of a passage, the subject then had to make educated predictions of the correct nth letter. From the resulting data, Shannon obtained experimental upper and lower bounds on the entropy of printed English. We reproduce his estimates in Table 2. Shannon's experiments were repeated by

Table 2

Experimental bounds for the entropy of English

n	Lower bound	Upper bound
1	3.19	4.03
2	2.5	3.42
3	2.1	3.0
4	1.7	2.6
5	1.7	2.7
6	1.3	2.2
7	1.8	2.8
8	1.0	1.8
9	1.0	1.9
10	1.0	2.1
11	1.3	2.2
12	1.3	2.3
13	1.2	2.1
14	0.9	1.7
15	1.2	2.1
100	0.6	1.3

Burton and Licklider (1955) using a much wider selection of literature. The results were roughly comparable.

While the estimates for these probabilities and entropies are reasonably accurate for most purposes, we should emphasize that they only apply to normal English prose. This would not, for example, include the novel 'Gadsby' by E. V. Wright, which in 250 pages never uses the letter 'e'.

Exercises 6.2

1. Taking the entropy of English as 1.5 bits, estimate the number of distinct meaningful strings of N symbols in English.

2. If you assume $H_E = 1.2$ bits show, by assuming the noiseless coding theorem, that 100 letters of ordinary text can be encoded in ~25.2 characters of recoded text without loss of information.

6.3 Zipf's law and word entropy

An alternative approach to the problem of finding the entropy of English, suggested by Shannon (1951), is to base the estimate on word frequency rather than letter frequency.

If English is regarded as a finite language having words $w_1, \ldots, w_N$ which occur independently with probabilities $p_1, \ldots, p_N$ respectively, then the *word entropy* H_W is given by

(1)
$$H_W = - \sum_{i=1}^{N} p_i \log p_i.$$

Shannon suggested that the symbol entropy H_E could be approximated from (1) by

(2)
$$H_E = H_W / \bar{w}$$

where $\bar{w}$ denotes the average length of a word in English.

There are various objections to such an approach; to name but two:

(a) The words used in English are not independent, and the formula (1) more correctly measures the *first-order word entropy*.

(b) Because the letters occurring in words are not independent, the relation (2) is a very crude approximation and, as we see below, is best replaced by an inequality.

Nevertheless, having made these points, it is informative to proceed with this approximation.

In order to evaluate H_W as defined by (1), we make use of a law proposed by the linguist G. K. Zipf (1935). This states that, if the words of any natural language are ordered in decreasing order of their probabilities of occurrence, so that p_n denotes the probability of the nth most probable word, then a good approximation to these probabilities is given by the formula

(3)
$$p_n = A/n,$$

where A is a constant that depends on the language in question.

Although Zipf's data has been criticized, it is striking that his law appears to hold well for languages as varied as Yiddish, Old German, Plains Cree, and Norwegian. For a reasoned exposition of the virtues and drawbacks of Zipf's law, and a discussion of possible explanations for it or some of its suggested extensions, we refer to Pierce (1962) or Miller (1981).

Shannon used Zipf's law as an approximation to the word frequencies of English, with the constant $A = 0.1$. The most frequent English word THE is plotted against 1; the next most frequent is OF and is plotted against 2, and so on, with AND and TO as the next most popular words. The fit to Zipf's law is shown in Fig. 1.

Using Zipf's law with $A = 0.1$ and taking $M = 12\,366$, we have

$$\sum_{n=1}^{12\,366} p_n = 0.1 \sum_{n=1}^{12\,366} \frac{1}{n} = 1,$$

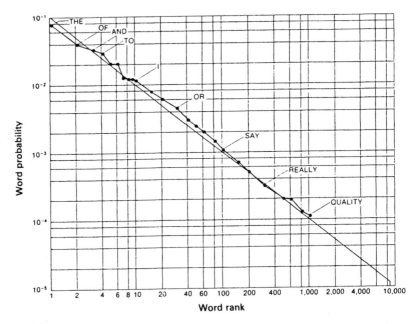

Fig. 1 *Illustration of Zipf's law for English.*

and then (1) gives

(4)
$$H_W = 9.72 \text{ bits per word.}$$

This is lower than the original estimate of Shannon which contained an arithmetical slip (see Yavuz, 1974).

Using the value (4) and taking the fairly well established approximation of 4.5 letters for the average length $\bar{w}$ of an English word, we obtain an estimate for $H_{4.5}$ (the *4.5 th-order approximation* to English in a 26-letter alphabet) of

(5)
$$H_{4.5} \simeq 9.72/4.5 = 2.16 \text{ bits/letter.}$$

Shannon realized that the estimate (5) is an underestimate. This is because the words in English are dependent sequences of letters; hence

$$H_W = \sum_{k=1}^{\infty} H(W \mid |W| = k) P(|W| = k),$$

where W is a 'random' word output and $|W|$ denotes the length (or number of letters) of W. Thus,

$$H_W = \sum_{k=1}^{\infty} H(X_1 X_2 \ldots X_k) P(|W| = k)$$

$$\leq \sum_{k=1}^{\infty} k H(X) P(|W| = k),$$

where $H(X)$ is the symbol entropy H_E and the inequality follows from the basic inequality

$$H(U, V) \leq H(U) + H(V).$$

This gives $H_W \leq H_{4.5} \sum k P(|W| = k)$, that is,

(6)
$$H_W \leq H_{4.5} \bar{w}.$$

Exercise 6.3

1. B. Mandelbrot (1954) suggested a modification of Zipf's law by replacing it with a rule

$$p_n = A/(n + V)^{1/D}$$

where D and V are independent parameters, and A is a constant to be fixed by the requirements that the total probability must sum to one. What values of D demand a finite vocabulary?

(For a discussion of the role of D as a fractal dimension, see Mandelbrot, 1977.)

6.4   The redundancy of a language

In our study of natural languages we have related the entropy per symbol of the language with the number of meaningful messages. More precisely, if Σ is the alphabet and $T(n)$ denotes the number of 'meaningful' or 'typical' messages of length n, then approximately we have

$$T(n) \approx 2^{nH}$$

where H is the entropy of the language.

Now, by the noiseless coding theorem, a source with entropy H has a compact encoding in an alphabet Σ in which the average length $\ell(n)$ of a typical string of n symbols satisfies

(1)
$$\ell(n) \simeq nH/\log |\Sigma| .$$

Hence, if we think of the redundancy R as a percentage, it is natural to define it by demanding

(2)
$$\ell(n) \simeq n(1 - R/100).$$

But combining (1) and (2) explains why Shannon was led to define the *redundancy* by

(3)
$$R = 1 - H/\log_2 |\Sigma|.$$

(By convention we often speak of a language as being, say, 30% redundant when $R = 0.3$.)

Estimating redundancy precisely is difficult; estimates obviously vary depending on texts chosen. Shannon, in his original work (1951), seems to have drawn his samples from *Jefferson the Virginian* by Dumas Malone. Newman and Waugh (1960) carried out a study of three sources, and some of their results are shown in Table 1, where H_n is their estimate of entropy based on n-grams.

Table 1

	The Bible	William James	The Atlantic Monthly
H_1	4.086	4.121	4.152
H_{12}	2.397	2.654	2.824
Estimates of redundancy	41.3%	32.2%	28.5%
Average word length	4.06	4.556	4.653

Not surprisingly, this quantifies the observation that the Bible represents the simplest prose and the Atlantic Monthly the most difficult.

More interesting is their study of identical passages of the Bible translated into different languages.

While Samoan is a language with only 16 letters, of which 60% are vowels, pre-1917 Russian used a 35-letter alphabet. Comparative results are shown in Table 2.

Table 2

	Samoan	English	Russian
H_1 (letter)	3.37	4.114	4.612
H_{12}	2.136	2.397	2.395
Redundancy	37.2%	41.3%	47.4%
Average word length	3.174	4.060	5.296

Shannon (1951) estimated that, taking into account the long-range effects, the entropy of English can be reduced to something of the order of one bit per letter, which would give a corresponding redundancy of roughly 75%. Although this figure was confirmed by Burton and Licklider (1955), it needs to be interpreted with care. For example, it does *not* mean that it is possible to resurrect a passage in which letters are deleted with probability $\frac{3}{4}$. The exact nature of the deletion is important.

If letters are randomly deleted with probability 0.5, then, for example, the message

MATHEMATICS IS BEAUTIFUL

might be distorted to

MTMASSBUFL;

thus it would be very difficult for the message to be resurrected purely from the distorted text. Indeed, studies of Chapanis (1954), confirmed by Miller and Friedman (1957), show that there is a critical value $p \simeq 0.25$ of the deletion probability above which recovery of the message from the mutilated text is impossible.

In other words, although it is perhaps theoretically possible to shorten printed texts to a quarter of their present length, random abbreviation does not seem a profitable way to achieve it.

However, it is clear that quite a big reduction can be achieved by 'sensible' encoding. For example, most readers would accept after a few trials the truth of the following statements:

(4)   It is usually possible to recreate passages of English in which every other letter (including spaces) is omitted.

(5)   It is possible to leave out all vowels from most pieces of prose and with some effort a reader can resurrect the text.

Both of these are examples of suboptimal encodings and, if we take the redundancy of 'average English' to be between 75% and 50%, then, because of (3), the entropy H_E satisfies

(6)
$$1.19 \le H_E \le 2.38.$$

Despite its rather nebulous and imprecise nature, the concept of redundancy is of crucial importance in cryptography, as we show in the next chapter.

Exercises 1. Alternate symbols (including spaces –) of the following sentence have
6.4   been omitted

MTEAISI – – RTYEATSINE.

What was the original sentence?
2. All vowels and spaces have been omitted from the following sentence

FCTSSTRNGRTHNFCTN

what was the original sentence?
3. Use the table in Appendix 2 to show that eliminating all vowels and spaces from English produces an average abbreviation of 48%.

PROBLEMS 6

1. Show that, in the zeroth-order approximation to English using the 27-letter alphabet, the probability that a string of 27 symbols contains no space is 0.361.
2. If it is assumed that the average length of a word in the first-order approximation to English is 4.5 letters, what is the probability of a space?
3. Prove that the average length of a word in the second-order approximation to English is the same as in the first-order approximation.
4. Let Σ be the 27-letter alphabet and let V be the subset of Σ consisting of the space and all the vowels; also, let C be the set of consonants and let

$$P_V = \sum_{i \in V} p_i.$$

Show that, assuming we can leave out all symbols from V and still recognize a message, then

$$H_E \leq \log(1 - P_V) - \left(\sum_{i \in C} p_i \log p_i\right) \Big/ (1 - P_V).$$

5. Consider a model of English reduced to two symbols (vowels and consonants). Show that, if we assume English is a Markov source, then the vowel/consonant model is also a Markov source.

6. Suppose a sample of prose from a language obeys Zipf's law exactly, and that the number of words in the language equals the number of occurrences of the most frequently used word. Show that, on average, a number of words equal to the square root of the number of different words in the passage will make up half of all the words in the sample of prose.

7. Is it true that a language in which each symbol is equally likely has zero redundancy?

8. Two languages obey Zipf's law exactly, but the first has twice as many words as the second. Show that, if q_1 and p_1 are the probabilities of the most frequent words in the two languages, then

$$p_1 \approx q_1/(1 - q_1 \ln 2).$$

9. Let Σ denote the 27-letter alphabet and let

$$q_1^N = \sum p(i_1, i_2, \ldots, i_{N-1}, j)$$

where the sum is over all $(N-1)$-grams $(i_1, \ldots, i_{N-1})$ and where j is defined to be the i that maximizes $p(i_1, \ldots, i_{N-1}, i)$. In other words, i is the symbol that would be next guessed by an 'ideal predictor' when shown the preceding $N-1$ symbols.

Similarly, q_2^N is the same expression, but with j taken to be the second most likely symbol guessed by the ideal predictor, and so on, until q_{27}^N is the probability associated with the least-likely symbol.

Prove that, for any G with $1 \leq G \leq 27$,

$$\sum_{i=1}^{G} q_i^N \geq \sum_{i=1}^{G} q_i^{N-1}.$$

(This is intuitively obvious, in that it expresses the fact that the probability of being correct about the next symbol in G guesses when given the N preceding symbols, is not less than when given only the preceding $N-1$ symbols.)

10. With the same notation as in the previous question, show that if H_N^E

denotes the N-gram entropy of English then

$$\sum_{i=1}^{27} i(q_i^N - q_{i+1}^N)\log i \le H_N^E \le -\sum_{i=1}^{27} q_i^N \log q_i^N.$$

(Shannon, 1951).
(These bounds for entropy in terms of these ideal-predictor probabilities were the basis of Shannon's derivation of entropy bounds as given in Section 2.)

7

Cryptosystems

7.1 Basic principles

Coded messages have a long history. Aeneas Tacticus includes a fascinating chapter entitled 'On secret messages' in his classic military manual on the art of war (dated c. 360 BC). In this chapter, Aeneas describes various devices, some of which might now appear rather primitive. His 'most secret method'—one which he was clearly proud of—was based on making holes in a sufficiently large astragal (a knucklebone of a sheep) and then drawing a thread through the holes of the astragal in an order corresponding to the letters of the message being sent.

A cleverer, or at least more secure method was used by the Spartans as early as 400 BC. They employed a device known as a skytale, which consisted of a tapered baton around which was wrapped in a spiral a strip of parchment containing the message. When unwrapped, the letters of the message appeared scrambled; however, when the parchment was wrapped around another baton of similar size, the original message reappeared.

The basic idea has remained constant: modify the message so as to make it unintelligible to anyone but the intended recipient. Typically, we represent a message M by a finite string of symbols from a finite alphabet Σ. We use $e(M)$ to denote the *encryption* of M and let d be the *decrypting* function satisfying the fundamental relationship

$$d(e(M)) = M$$

for all messages M.

In practice, e can be regarded as a function or algorithm with a collection of parameters. Any such parameter is called a *key*, and is usually denoted by K. The encrypted string

$$C = e(M, K)$$

is called the *cipher*, *ciphertext*, or *cryptogram*; again, decryption obeys

$$d(C, K) = M.$$

The message M is often called the *plaintext*.

Thus, formally, we define a *cryptosystem* to be a triple $\langle M, K, C \rangle$, where M and C are sets (usually of strings from finite alphabets Σ_1 and Σ_2 respectively) and K is the finite set of keys with the additional hypothesis that there exist functions (or algorithms) e and d such that

$$e: M \times K \to C, \qquad d: C \times K \to M,$$

and, for each $(M, K) \in M \times K$,

(1)
$$d(e(M, K), K) = M.$$

We now describe several well known examples of cryptosystems. Throughout, Σ will denote the language of the system, which we usually take to be the letters of the English alphabet, sometimes augmented by a 'blank' or 'space'.

SIMPLE SUBSTITUTION

In this system, the key is permutation π of Σ and each letter of the message is replaced by its image under the permutation π. Usually, the key is represented as a 26-letter string, say UXEB ..., and then any occurrence of A in the message is replaced by U, B by X, C by E, D by B, and so on. Spaces are ignored and usually left out in the cryptogram.

A substitution cipher is sometimes described as a *monoalphabetic* cipher. Such systems are extremely vulnerable to attack by a frequency analysis of letters, digrams, etc., and a first requisite of a 'safe' system is that it be *polyalphabetic,* so that the encipherment of a specific symbol changes as the plaintext is being encoded.

N-GRAM SUBSTITUTION

Instead of substituting for letters, one can substitute for digrams, trigrams, etc. General digram substitution, for example, will require a key consisting of a permutation of the 26^2 different digrams, and can best be represented by a 26×26 standard array in which the rows correspond to the first letter of the digram and the columns to the second letter, entries in the array being the substitutes.

TRANSPOSITION OF ORDER *d*

For any positive integer d, divide the message M into blocks of length d. Then take a permutation π of $1, 2, \ldots, d$ and apply π to each block. For example if $d = 5$ and $\pi = (4\,1\,3\,2\,5)$, a message such as

$$M \equiv \text{JOHN} \mid \text{IS A} \mid \text{GOOD} \mid \text{SKIER}$$

is transformed into

$$C = \text{ONHJ} \mid \text{SA I} \mid \text{ODOG} \mid \text{KEISR}$$

Decryption is carried out using the inverse permutation π^{-1}.

Transposition has an advantage over substitution in that, although it preserves the frequency distribution of single letters, it destroys the digram, trigram, and higher-order statistics of the language. For this reason, it is a safer method of enciphering a natural language than simple substitution.

THE VIGENÈRE AND CAESAR CIPHERS

In this system, the key consists of a series of d letters. These are written repeatedly below the message and added modulo 26, considering the alphabet as numbered from $A = 0$ to $Z = 25$.

For example, with $d = 3$, if the key consists of the sequence ABC then the message

$$M \equiv \text{JOHN IS GOOD}$$

with

$$K \equiv \text{ABCABCABCA}$$

becomes

$$C \equiv \text{JPJNJUGPQD}.$$

The special case where $d = 1$, so that the key is just a single letter, is called the *Caesar cipher* and was allegedly used by Julius Caesar. It is particularly simple to crack but is a useful vehicle for explaining cryptographic principles.

THE PLAYFAIR CIPHER

This is a scheme based on digram substitution where the key is a 5×5 square consisting of some arrangement of the 25 letters of the alphabet (the letter J being omitted because of its infrequent use and easy representation as an I). Suppose, for example, the key square is

$$\begin{array}{ccccc}
B & Y & D & G & Z \\
W & S & F & U & P \\
L & A & R & K & X \\
C & O & I & V & E \\
Q & N & M & H & T
\end{array}$$

The substitute for a digram AV (say) is the pair of letters at the other corners of the rectangle defined by A and V, that is, OK; the O is taken first since it is below A. If the digram letters are on a horizontal line as for instance, WU, then one uses the pair of letters on their right, in this case SP. If the digram letters are on the same vertical line, for example OS, then they are replaced by the letters immediately below them, in this case NA.

As might be expected, the key square is regarded as toroidal so that, for example in the key above, ON is transposed into NY and UP is transposed into PW. Also, special rules have to be made for the case of repeated letters, such as omitting the second letter or separating them by blanks and so on. This can cause problems, for example when encoding LESS SEVEN. There are many variants of the original Playfair scheme, we refer the reader to M. V. Gaines (1956).

THE AUTOKEY CIPHER

In this system, there is a 'priming' key, which is usually short and is used to start the encipherment. The enciphering process is then continued, using either the message itself or the running cryptogram. Consider the following example from Gaines (1956): Suppose that the priming key is COMET and the message is SEND SUPPLIES; then, using the message as a running key, we encode to get using the obvious addition mod 26.

M:	S E N D S U P P L I E S
K:	C O M E T S E N D S U P
Cryptogram	U S Z H L M T C O A Y H

Using the cryptogram as key, the same priming key gives

M:	S E N D S U P P L I E S
K:	C O M E T U S Z H L O H
Cryptogram	U S Z H L O H O S T S Z

For a discussion of how to crack this system, we refer to Gaines or Givierge (1925) who attributes it to Commandant Bassières.

LINEAR TRANSFORMATIONS

Two of the earliest papers to treat cryptography as a piece of formal mathematics were by L. S. Hill in 1929 and 1931.

The basic idea was to split the message into blocks of length d and to identify such a block with a d-tuple x of integers, and then multiply this vector x by a nonsingular matrix U. The cryptogram would then be the vector $y = Ux$ and decryption is carried out by multiplying y by the inverse matrix U^{-1}.

To ensure that the arithmetic (algebra) is carried out over a field, all calculations are modulo a prime integer. Typically the letters A–Z are represented by the integers 0–25, the space by 26, comma 27, and full stop 28, giving an alphabet of 29 symbols. Alternatively the alpha-numeric alphabet Σ of size 37 is used by using 0–9 for the decimal integers, blank by 10, and the 26 letters of the Roman alphabet by 11–36.

Example   Working with the 37-letter alphabet and with block size $d = 2$, if the key is the matrix.

$$U = \begin{bmatrix} 3 & 13 \\ 22 & 15 \end{bmatrix},$$

the message GOOD is first written as

$$GOOD \equiv 17, 25, 25, 14.$$

Then, since

$$\begin{bmatrix} 3 & 13 \\ 22 & 15 \end{bmatrix}\begin{bmatrix} 17 & 25 \\ 25 & 14 \end{bmatrix} = \begin{bmatrix} 6 & 35 \\ 9 & 20 \end{bmatrix}$$

in the field of 37 elements, the cryptogram is

$$C \equiv 6, 9, 35, 20. \qquad \square$$

A basic distinction in some of the literature on cryptography is between stream and block ciphers. The plaintext message is a sequence of symbols from the alphabet Σ (typically letters, numbers or more often binary digits). A *stream cipher* operates on the message, symbol by symbol, to produce the cryptogram. In a *block cipher* the message is broken up into blocks and the encryption algorithm then operates on each block separately.

Although this distinction between block and stream ciphers can be useful, there is no real mathematical difference between the two modes. To see this, note that, if the lengths of the blocks are say n, then a block cipher is just a stream cipher on the enlarged alphabet $\Sigma^{(n)}$. Because of this, we will not labour the distinction henceforth.

Exercises
7.1

1. What would be the effect on the cryptosystem based on the method of linear transformations if the transforming matrix U was singular?
2. How many keys does a Playfair system have?

7.2 Breaking a cryptosystem

What does it mean to break a cipher system? The *cryptanalyst,* or *enemy* as he is commonly called, is assumed to have full knowledge of the encrypting function e and the decrypting function d. In addition, he may also have a variety of side information such as language statistics, knowledge of context, etc.

He will certainly have some cipher text C, and all he lacks is the key K from which he can use d to decipher C correctly. The situation may be represented pictorially as in Fig. 1.

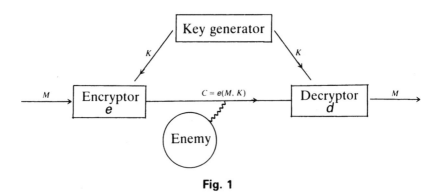

Fig. 1

There are three possible attacks which we envisage an enemy making on the system:

(a) *Cryptogram-only attack (or ciphertext-only attack):* this is the situation described above where the enemy only knows some piece of ciphertext C.
(b) *Known-plaintext attack:* In this case—a more realistic one than

(a)—the enemy is assumed to possess a considerable length of
message text and corresponding ciphertext, and from this seeks
to find the key.

This is a much more formidable attack to withstand but is
nowadays regarded as a bare minimum security standard. Indeed,
Lempel (1979, p. 290) asserts that, at the present time, the most
appropriate criterion for judging a secrecy system is that of its ability
to withstand the following even stronger attack.

(c) *Chosen-plaintext attack*: In this case, the enemy is able to acquire
an arbitrary number of corresponding message and cipher pairs
(*M, C*) of his own choosing.

It is not difficult to see that none of the secrecy systems described
in the previous section can withstand a chosen-plaintext attack.

The situation concerning their vulnerability to known-plaintext
attack is more complicated. Consider, for example, the simplest
possible system—simple substitution—so that the key K is a
permutation of the 26 letters in which (say) A $\rightarrow$ P, B $\rightarrow$ F, and so on.
Now, if the message sent is the 'silly message' consisting of the letter
A repeated n times, then the corresponding cryptogram C is the
letter P repeated n times. No matter how large an n we have, this
particular message–cryptogram pair will not yield the key K.
However, it is not hard to believe that, on average, given a
reasonable length of message and cryptogram, any of the above
systems can be cracked. These crude notions of 'on average' and
'reasonable length' will be clarified in the next few sections.

Exercises　1. Show that the linear-transformation cryptosystem of Hill is vulnerable to
7.2　　　chosen-plaintext attack. What is the minimum length of
　　　　plaintext/ciphertext needed to crack a system which uses a block of length
　　　　d?

　　　　2. If you are mounting the most efficient (that is, shortest) chosen-message
　　　　attack on a cryptosystem based on (a) simple substitution, (b) Vigenère of
　　　　order d, which message would you use?

7.3 Equivocation and perfect secrecy

Suppose that we have a cryptosystem $\langle M, K, C \rangle$ and suppose further
that the set M of possible messages is finite and that we may assume:
(a) p_i is the probability that the message sent is M_i, for $1 \le i \le n$;
(b) the probability that the key used is K_j is given by k_j, and the

choice of key is independent of the message being transmitted.

These two probability distributions induce a probability distribution on the set of possible cryptograms, where for any particular C, say C_u, the probability that the 'random' cryptogram C equals C_u is given by

$$P(C = C_u) = \sum p_i k_j,$$

where the sum on the right is over all message–key pairs (M_i, K_j) such that $e(M_i, K_j) = C_u$.

The situation now is completely analogous to a communication channel as described in Chapter 3. We can regard M as a zero-memory source with the encrypting function e, together with the keys serving it as the channel.

The critical concept is what Shannon called the *key equivocation* $H(K \mid C)$. It measures the average uncertainty remaining about the key when a cryptogram has been intercepted. In the same vein we define the *message equivocation* to be $H(M \mid C)$. Sometimes we write S for a cryptosystem $\langle M, K, C \rangle$ and then denote by $H(S)$ the equivocation $H(K \mid C)$.

A variable message M has an entropy

$$H(M) = - \sum p_i \log p_i,$$

and, if K and C denote respectively a variable key and variable cryptogram, then the entropies $H(K)$ and $H(C)$ are defined as expected by

$$H(K) = - \sum P(K = K_i) \log P(K = K_i),$$

$$H(C) = - \sum P(C = C_j) \log P(C = C_j),$$

where the summations run through the possible keys K_i and cryptograms C_j.

The following property of equivocation expresses mathematically the fact that there is more uncertainty associated with the key than with the message.

Theorem 1   *The key equivocation is related to the message equivocation by*

(1)
$$H(K \mid C) = H(M \mid C) + H(K \mid M, C).$$

Corollary   *The key equivocation is at least as big as the message equivocation.*

Proof   This follows from repeated use of the fundamental identity

$$H(X \mid Y) = H(X, Y) - H(Y).$$

For we can write

(2)
$$\begin{aligned} H(M \mid C) &= H(M, C) - H(C) \\ &= H(M, K, C) - H(K \mid M, C) - H(C). \end{aligned}$$

Now

$$\begin{aligned} H(K \mid C) &= H(K, C) - H(C) \\ &= H(M, K, C) - H(M \mid K, C) - H(C). \end{aligned}$$

But

$$H(M \mid K, C) = 0$$

since, once a cryptogram C and K are known, the message sent is uniquely determined. Thus

$$H(K \mid C) = H(M, K, C) - H(C)$$

which, when compared with (2), proves the identity (1).   □

Since $H(M \mid C)$ measures the average uncertainty of the message after interception of the cryptogram, it is natural to say that a system $\langle M, K, C \rangle$ has *perfect secrecy* if $H(M \mid C) = H(M)$.

As we shall see, it is difficult to find practical examples of such systems.

Theorem 2   *A system $\langle M, K, C \rangle$ has perfect secrecy if and only if, for all possible messages $M \in M$ and possible cryptograms $C \in C$, we have*

(3)
$$P(M \mid C) = P(M).$$

Note 1   In the above statement $P(M \mid C)$ has its obvious interpretation as the probability that a specific message M is sent, given that the cryptogram C is received.

Note 2   In the statement of Theorem 2, we are assuming that the adjective 'possible' implies that the message (cryptogram) has a nonzero probability.

Proof   We know from (1.3.2) that $H(X \mid Y) = H(X)$ if and only if X and Y

are independent random vectors. But this is exactly the condition (3). □

An almost immediate consequence of Theorem 2 is:

Corollary *A cryptosystem $\langle M, K, C \rangle$ is perfect if and only if, for any possible message M and cryptogram C, we have*

(4) $$P(C \mid M) = P(C).$$

Theorem 3 *A necessary condition that a cryptosystem has perfect secrecy is that it has at least as many keys as messages.*

Proof Let there be n possible messages. Consider any fixed key K. Then it must map the n distinct messages $M_1, \ldots, M_n$ into distinct cryptograms which, without loss of generality, we call $C_1, \ldots, C_n$. Hence

$$P(C_j) = P(C_j \mid M_j) > 0$$

for any C_j $(1 \le j \le n)$. But the system is perfect and hence, because of (4), for any other message M_u, with $u \ne j$, we have

$$P(C_j \mid M_u) = P(C_j) > 0.$$

Hence there must be some other key K' mapping M_u to C_j. This must hold for each u $(1 \le u \le n)$ with $u \ne j$, and all such keys must be distinct. Hence there must be at least n keys. □

As might be expected, most contemporary cryptosystems are unlikely to achieve this level of security. We return to this in §8.1.

Exercises 7.3
1. Show that, if d is larger than some integer d_0 (which should be found), then the system of transposition or order d has greater key equivocation than that of the Vigenère system of order d.
2. Prove, for any cryptosystem $\langle M, K, C \rangle$, the identities

$$H(C \mid M) = H(M, K, C) - H(M) - H(K \mid C, M)$$
$$= H(K) - H(K \mid C, M).$$

7.4 Combining cryptosystems

A natural way of trying to increase security is to take different systems and combine them in various ways. Two such methods, which were suggested by Shannon (1949), still form the basis of many practical cryptosystems. They are:

(a) *The weighted sum*

If S_1 and S_2 are two cryptosystems with the same message space (domain), and $0 < p < 1$, then their *weighted sum* $pS_1 + (1-p)S_2$ is the cryptosystem determined by making a preliminary choice: using S_1 with probability p, or S_2 with probability $1-p$.

Thus, if S_1 has keys $K_1, \ldots, K_m$ with p_i the probability of using K_i, and S_2 has keys $K'_1, \ldots, K'_n$ with p'_i the probability of using K'_i, then the system $pS_1 + (1-p)S_2$ has $m+n$ keys $K_1, \ldots, K_m,$ $K'_1, \ldots, K'_n$, where K_i is used with probability pp_i and K'_j is used with probability $(1-p)p'_j$.

This has a natural extension to more than two systems.

(b) *The product*

A second way of combining systems S_1 and S_2 is by first applying S_1 to our message and then applying S_2 to the resulting cryptogram. Obviously, for this to be well defined, the domain of S_2 must contain the range of S_1. When it exists, we denote this product by $S_1 * S_2$.

If the keys of S_1 are chosen with probabilities $p_1, \ldots, p_m$ and those of S_2 with probabilities $p'_1, \ldots, p'_n$, then $S_3 = S_1 * S_2$ has mn keys with probabilities $p_i p'_j$. Notice that $S_1 * S_2$ may often have fewer than mn effective keys, since some of the product transformations may agree.

Algebraically, we can sum up the properties of combinations of cryptosystems in the following remarks:

If S_1, S_2, S_3 are cryptosystems between which the relevant operation is well defined, and q is $1-p$, then

(1) $$S_3 * (pS_1 + qS_2) = pS_3 * S_1 + qS_3 * S_2,$$

(2) $$(pS_1 + qS_2) * S_3 = pS_1 * S_3 + qS_2 * S_3,$$

(3) $$S_1 * (S_2 * S_3) = (S_1 * S_2) * S_3,$$

(4) $$S_1 * S_2 \text{ does not generally equal } S_2 * S_1.$$

Exercises 7.4

1. If S_1 and S_2 are both transposition systems of order d, show that their product is well defined but that $S_1 * S_2$ has only the same number of effective keys as S_1.

2. If S_1 is a transposition system of order d_1, and S_2 is a transposition system of order d_2, show that $S_1 * S_2$ is well defined. How many keys does it have?

3. If S_1 and S_2 are cryptosystems with key equivocations H_1 and H_2 respectively, assuming that their weighted sum is well defined, find the key equivocation of $pS_1 + (1-p)S_2$.

7.5 Unicity

Most readers would have little confidence that any of the codes described in the previous sections could withstand any sustained attack. In this section we will try to formalize the reasons for this vulnerability.

We first illustrate the basic idea by an example.

Example Suppose a cryptosystem is based on the Caesar cipher with the 27-letter alphabet, so that

$$A = 0, \quad B = 1, \ldots, Z = 25, \quad \text{space} = 26,$$

and addition is modulo 27. You intercept the cipher text

(1) HWXGOQCCZOXGOXBORCST.

We propose the following attack to find the unknown key K_0. Choose candidate keys for K_0 at random and invert the cipher text, symbol by symbol, for as long as the message got is 'meaningful'.

Thus, if our first choice of key is $K_0 = E$, we will invert (1) to get

DSTC ... ,

at which stage we declare E not to be the key K_0. Choose again (say) $K_0 = Z$, which inverts (1) to

JYZ ... ,

which rules out Z.

Our next (? inspired) choice for K_0 is P and is quickly seen to be the correct key transforming the cipher text (1) into a meaningful message

THIS BOOK IS IN CODE.

At this stage we have no guarantee that there is no other message–key pair that would have given the same cipher text. However routine testing of other keys shows this to be so. □

The success of this method of attack hinges on the fact that there is *only one* message–key pair from which the cipher text could have arisen. Had our ciphertext contained only two symbols, this would clearly not have been true. The *unicity* of a given cryptosystem can be loosely defined to be the average length of ciphertext such that there is precisely one plaintext–key pair from which the ciphertext could have arisen. However, there are considerable problems in this definition, particularly surrounding the word 'average'.

Accordingly we adopt a slightly different approach with essentially the same results. Given a cryptosystem $\langle M, K, C \rangle$ we let M_N and C_N denote 'random' pieces of plaintext and corresponding ciphertext of length N.

Now, in an obvious notation,

$$
\begin{aligned}
H(K \mid C_N) &= H(K, C_N) - H(C_N) \\
&= H(M_N, K, C_N) - H(C_N) \\
&= H(M_N, K) - H(C_N) \\
&= H(M_N) + H(K) - H(C_N).
\end{aligned}
$$

Thus we *define* the *unicity* or *unicity point* U to be the least value of $N > 0$, if it exists, such that $H(K \mid C_N) = 0$, and therefore the value of N such that

(2)
$$
H(M_N) + H(K) - H(C_N) = 0.
$$

Suppose now we make the following assumptions:
(a) the underlying (natural) language is such that a reasonable estimate of $H(M_N)$ is given by

(3)
$$
H(M_N) \simeq NH,
$$

where H is the entropy per symbol of the language;
(b) the cryptosystem is such that all N-sequences of symbols are equally likely as cipher texts; in other words,

(4)
$$
H(C_N) \simeq N \log |\Sigma|.
$$

(This is not unreasonable: any good cryptosystem should aim at possessing this property.)

Accordingly, substituting these values into (2) means that, if the system satisfies (a) and (b), then U satisfies

$$
UH + H(K) - U \log |\Sigma| = 0,
$$

that is,

(5)
$$
U = \frac{H(K)}{\log |\Sigma| - H}.
$$

Usually we assume that all keys are equally likely to be chosen, as this means that

(6)
$$
U = \frac{\log |K|}{\log |\Sigma| - H},
$$

where H is the entropy per symbol of the source.

UNICITY AND REDUNDANCY

There is a close relationship between the unicity point of a cryptosystem and the redundancy of the language being transmitted. Recall that the redundancy R of a language with entropy H is given by

$$R = 1 - \frac{H}{\log |\Sigma|}$$

so that

$$U = \frac{\log |K|}{\log |\Sigma| - H} = \frac{\log |K|}{R \log |\Sigma|}.$$

Thus we have a precise statement of what (when you think about it) is obvious, namely:

(7)   If a language has zero redundancy then, for any cryptosystem satisfying (a) and (b), the unicity point is at infinity.

Example Suppose we take the simple substitution cipher so that there are 26! keys. Taking $\log 26 = 4.7$ and the entropy H_E of English to be 2 bits per symbol we get from (6)

$$U = \frac{\log 26!}{4.7 - 2} = \frac{88.4}{2.7} \approx 32. \qquad \square$$

In other words, taking the entropy of English to be 2 bits/symbol (which is probably a little high) we have obtained a unicity value of 32 symbols. This is in fairly good agreement with the following empirical observation of Shannon (1949), who claimed that the unicity point

'can be shown experimentally to lie between the limits 20 and 30. With 30 letters there is nearly always a unique solution to a cryptogram of this type and with 20 it is usually easy to find a number of solutions'.

In a similar vein Friedman (1973) claims that

'Practically every example of 25 or more characters representing monoalphabetic encipherment of a "sensible" message in English can be readily solved.'

Exercises 7.5 1. Show that, if we take the unicity point of the substitution cipher to be at least 25 symbols, then this gives a lower bound of 1.4 bits/symbol for the entropy of English.

2. Explain why the transposition ciphers do not satisfy (4). How should the unicity formula (5) be modified?

(Hellman, 1977; Shannon, 1951)

3. Prove that $H(K \mid C_N)$, where C_N denotes a ciphertext of length N, is a monotone decreasing function of N.

7.6 Hellman's extension of the Shannon theory

The classical information-theoretic approach to cryptography, as described in the previous sections, was pioneered by Shannon in his fundamental paper. In 1977, M. E. Hellman proposed an alternative and attractive extension of these principles. The following approach is based on Hellman's work.

In this model, the message space is divided into two disjoint subsets. The first contains 2^{HN} meaningful or typical messages, each with the same a priori probability 2^{-HN}. The remaining messages are meaningless in the language and are assigned probabilities of 0. In other words, Hellman is postulating that the message source has the AEP property of ergodic sources discussed in Section 5.4. We also assume the keys are used independently of the message and with equal probability.

Referring to Fig. 1, a line joining a message M_i to a cryptogram C_j is labelled with the index of the key encoding M_i to C_j. Thus, in the figure $e(M_1, K_7) = C_5$. There may be more than one labelled line

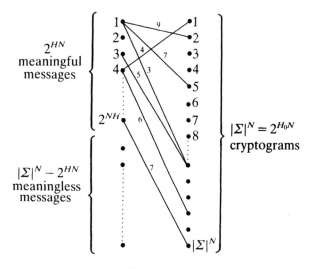

Fig. 1

joining a meaningful M_i to C_j, but all lines leaving any M_i will be labelled with different keys and all lines incident with a given C, will also be labelled with different keys. In the figure, we only show lines joining the cryptograms with the meaningful messages.

If C is any cryptogram, define $Z(C)$ to be the number of pairs (M, K) such that M is meaningful and $e(M, K) = C$.

Now consider a given cryptogram C. If there exists more than one pair (M_i, K_j) such that M_i is meaningful and

$$e(M_i, K_j) = C,$$

then the enemy, on intercepting C, will be left in some doubt about the key used. Thus, if $Z(C) > 1$, the cryptogram C will have a *spurious* key decipherment. Hence, if $s(C)$ is defined by

$$s(C) = \max\{[Z(C) - 1], 0\},$$

then the expected value of $s(C)$, namely

$$\bar{s} = \sum_{C \in C} s(C) P(C),$$

measures the expected number of spurious key decipherments. But it is easy to see that

(1)
$$\bar{s} = \bar{z} - 1,$$

where

(2)
$$\bar{z} = \sum_{C \in C} Z(C) P(C) = \sum_{C \in C} Z^2(C) / 2^{NH} |K|,$$

since, by definition of the model, for any cryptogram C,

$$P(C) = Z(C) / 2^{NH} |K|.$$

But, by a simple counting argument (see Fig. 1),

(3)
$$\sum_C Z(C) = 2^{NH} |K|.$$

We now use the elementary lemma that, for any x_i satisfying

$$\sum_{i=1}^{n} x_i = a,$$

we have

$$\sum_{1}^{n} x_i^2 \geq a^2 / n.$$

Hence, from (2) and (3), we get

$$\bar{z} \ge (2^{NH} |K|)^2 / |C| 2^{NH} |K|$$
$$= 2^{NH} |K| / |C|.$$

Combining this with (1) means that we have proved the following result.

Theorem   *Under the assumptions stated, the expected number of spurious key decipherments satisfies*

(4)
$$\bar{s} \ge (2^{NH} |K| / |C|) - 1.$$

If we now write

$$|K| = 2^{H(K)}, \qquad |C| = |\Sigma|^N = 2^{NH_0},$$

where H_0 is the zeroth-order entropy of the language, then (4) reduces to

(5)
$$\bar{s} \ge 2^{NH+H(K)-NH_0} - 1,$$

and the right-hand side equals zero precisely at the unicity point as defined in the previous section.

Example   Suppose that the cryptosystem is a Vigenère with key length 80, and that we are using the 26-letter English alphabet. With $H_0 = 4.7 = \log 26$ and

$$H = H_E \approx 1.5 \text{ bits}, \qquad H(K) = 80 \log 26 \approx 376,$$

there will be on average at least $2^{376-320} \approx 2^{56}$ different spurious key decipherments from a cryptogram of 100 letters.   □

7.7   Conclusion

The classical Shannon approach to cryptography as described in the preceding sections gives a good qualitative insight into the behaviour and design principles of cryptosystems. However, it should be emphasized that it is only a mathematical model. Some of the assumptions made are likely to be only approximately satisfied.

However, from the theory, we have obtained the main mathematical principles enunciated by Shannon. The first of these is that to achieve perfect secrecy the key space must be as large as the message space. In the next chapter we show how this is achieved.

The second principle of Shannon was that practical systems have to rely for their safety on the hope (?belief) that cracking the system would be too time-consuming and expensive to be practical. Shannon introduced the concept of a *work characterisic* of a cryptosystem, which he defined to be the average amount of work (measured in man-hours) needed to determine the key used from a cryptogram of a given size. Thus 'work characteristic' can be regarded as the forerunner of the 'computational complexity' of the cryptanalysis problem, which we discuss in Chapter 9. In Chapters 10–13 we will show how to make use of the theory developed to obtain practical, mathematically safe, systems.

PROBLEMS 7

1. Suppose that the message space consists of all n-digit decimal integers, and that encipherment is by a Caesar cipher in which the key K is a single decimal digit and addition is modulo 10. (For example, $K = 3$, $M = 258$, and $C = 581$.)

   On the assumption that all values of M and K are equally likely, find $H(M \mid C)$ and $H(K \mid C)$ and the unicity of the system.

2. Prove that, for any cryptosystem,

$$H(K, C) = H(M) + H(K).$$

3. How long a message is needed to break (a) the transposition of order d; (b) the Hill system of block length d, if the attack is by chosen plaintext.

4. Prove that

$$H(S_1 * S_2) \geq H(S_1).$$

   Show by example that it is not true in general that

$$H(S_1 * S_2) \geq H(S_2).$$

5. Prove that, if S_1 and S_2 have perfect secrecy, then so does $S_1 * S_2$.

6. Calculate the unicity of the Caesar cipher and reconcile the answer you get, based on taking the entropy of English as 1.5 bits per symbol, with the estimate you make from an experiment with various pieces of English text.

   Note: It is nontrivial to get two distinct pieces of meaningful English of length ≥ 6 which might be encoded with the same key. There are strings of length 17. One such pair was found by S. H. Babbage (1987).

7. Show that, in a perfect cryptosystem, $H(K \mid C) \equiv H(K)$.

8. Let S be a cryptosystem obtained by taking the convex combination of systems $S_1, \ldots, S_n$, so that

$$S = p_1 S_1 + \ldots + p_n S_n,$$

where $p_i \geq 0$ and $\sum p_i = 1$. Show that

$$\sum_{i=1}^{n} p_i H(S_i) \leq H(S) \leq \sum_{i=1}^{n} p_i H(S_i) - \sum_{i=1}^{n} p_i \log p_i,$$

where

$$H(S) = H(M \mid C).$$

(Shannon, 1949)

9. What are the bounds on the unicity for a cryptosystem consisting of a transposition of order d on the 26-letter English alphabet, if the entropy of English is taken as lying between 1 and 2 bits per symbol?

10. An *affine block* cryptosystem is one in which the key is a nonsingular square $d \times d$ matrix A together with a d-vector t. It works by breaking the message up into binary blocks of size d, then

$$C = AM + t$$

where M is a particular block of length d and all arithmetic is modulo 2.
    (a) Show that the number of keys is

$$2^d (2^d - 1)(2^d - 2)(2^d - 2^2) \ldots (2^d - 2^{d-1}).$$

    (b) Prove that the cryptosystem is vulnerable to chosen message attack, and find the minimum length of plaintext–ciphertext needed to find the key.

11. A cryptosystem works by breaking the message into binary blocks of size d and then encrypting M by

$$C = \pi(M) + t,$$

where π is a permutation and t is a d-vector, and all the arithmetic is binary.
    (a) How large is the key space?
    (b) Show that this cryptosystem is an affine block system.

12. An *affine Caesar substitution* is a 1–1 transformation of $\mathbb{Z}_N$ (the integers modulo N) into itself by a transformation of the form $x \mapsto E(x)$ where

$$E(x) = ax + b \bmod N \qquad (0 \leq x \leq N - 1)$$

and where a and b are positive integers less than N.
    Show that E is a $1 - 1$ transformation if and only if a and N are relatively prime. Hence find the number of affine Caesar substitutions on the 26-letter alphabet.

13. Consider the Vigenère cipher of order d. What is its unicity point when applied to English text of entropy 1.5 bits/symbol?

14. Show that, if we accept as true the empirical evidence that any ciphertext of 40 or more symbols from the 26-letter alphabet can be uniquely

deciphered into meaningful plaintext by a substitution cipher, then the entropy of English is ≤ 2.5 bits per symbol.

15. Prove that a cryptosystem $\langle M,\ K,\ C \rangle$ with $|M| = |K| = |C|$ has perfect secrecy if and only if:
    (a) there is exactly one key transforming each message to each cryptogram, and
    (b) all keys are equally likely.

16. Given a cryptosystem $\langle M,\ K,\ C \rangle$, let e_i denote the encryption function using key K_i and d_j the decryption function using key K_j. The system is called *pure* if, for any $i,\ j,\ k$, there exists l such that, for all $M \in M$,

$$e_k d_j e_i(M) = e_l(M).$$

The (message) *residue class* $R(M)$ is defined to be the set of messages M' such that, for some i and j,

$$d_j e_i(M) = M'.$$

Prove that, if the system is pure, the residue classes $R(M)$ partition M. Show also that, if $M_1 \in R(M)$, then

$$R(M_1) = R(M).$$

($R(M)$ is the set of messages which a cryptanalyst might confuse with M.)

17. With the same notation as in the previous question, define the *cryptogram residue* class $C(M)$ of a message M to be the set of cryptograms into which M could be encoded using all the keys of the system. Suppose that $\langle M,\ K,\ C \rangle$ is pure. For any message M, show that
    (a) the cryptogram residue classes partition C;
    (b) $|R(M)| = |C(M)|$ for all $M \in M$;
    (c) if M_1 and M_2 belong to the same message residue class, then $C(M_1) = C(M_2)$.

8

The one-time pad and linear shift-register sequences

8.1 The one-time pad

The classic example of a perfectly secure cryptosystem is the *one-time pad*. This was first introduced by G. S. Vernam in 1926. It works as follows. Suppose that the alphabet Σ is the usual 26-letter English alphabet, that punctuation, spaces, etc. are to be omitted, and that the message M to be sent contains N letters. To encipher M, generate a random sequence of N letters from the alphabet Σ, with each letter having probability $\frac{1}{26}$ of being chosen and each choice being made independently. This random sequence $(Z_1, \ldots, Z_N)$ will be the key K and, in order to encipher $M = (x_1, \ldots, x_N)$ using K, we just define

$$C = e(M, K) = (y_1, \ldots, y_N)$$

by

$$y_i = x_i + Z_i \pmod{26}$$

where, as usual in substitution cipher systems, we are giving letters a numerical value in the range $\{0, \ldots, 25\}$ corresponding to their order in the alphabet.

It is easy to see that, for this system, there are 26^N equally probable keys, so that

$$H(K) = N \log 26.$$

Since the key $(Z_1, \ldots, Z_N)$ is a sequence of independent symbols from Σ, it is also clear that there are 26^N equally likely cryptograms C and that

$$P(K \mid C) = 1/26^N$$

for any $K \in K$. From this, Theorem 7.3.2 gives us the next result.

Theorem 1 *The one-time pad has perfect secrecy.*

However, there are enormous drawbacks in using such a system. Among these are the following.

(a) There is no mathematical way of generating independent random variables to act as a key, and one would have to rely on a 'pseudorandom' sequence generated by one of the many standard methods. There is no guarantee that such pseudorandom sequences would give the same level of security. This is a deep mathematical problem which has recently attracted a lot of attention, and we return to it in the last chapter.

(b) The length of key is exactly the same as that of the message and, if a 'genuine random sequence' obtained by some physical process (such as radioactive decay) is used as a key, then this key must also be communicated to the receiver before decryption can be carried out.

Despite these drawbacks, it is allegedly used for the very highest levels of communication such as the Moscow–Washington hotline (Simmons, 1979, p. 316).

Exercises 8.1

1. A common method of generating 'pseudorandom' number sequences $(r_n : 1 \le n < \infty)$ is by the method of linear congruences: for each $n > 0$, define

$$r_{n+1} = ar_n + b \pmod{m}, \qquad r_0 = c,$$

with a, b, m, and c chosen initial integers (m is called the *modulus* and c the *seed*; clearly a, b, and c must be in the range $(0, m-1)$).
Show that

(a) if the pseudorandom sequence is to contain all integers in the range $[0, m-1]$, then $(r_1, \ldots, r_m)$ must be a permutation of $(0, \ldots, m-1)$.

(b) if b, c, and m are each divisible by p, then r_n is divisible by p for all n.

8.2 Linear shift-register sequences

A (linear) shift register with feedback is an arrangement of registers in a row, each register being capable of holding either the digit 1 (on) or 0 (off). A clock pulse regulates the behaviour of the system, which works according to the following laws.

Suppose that the system has n registers $R_1, \ldots, R_n$ as shown in Fig. 1 and that $X_i(t)$ denotes the content of register R_i at time t. Initially, let the system be given the configuration

$$x(0) = (X_1(0), \ldots, X_n(0)).$$

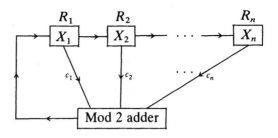

Fig. 1

If

$$x(t) = (X_1(t), \ldots, X_n(t))$$

denotes the state of the system at time t, the state at time $t+1$ is given by the rules

(1)
$$X_i(t+1) = X_{i-1}(t) \quad (2 \le i \le n),$$

(2)
$$X_1(t+1) = c_1 X_1(t) + \ldots + c_n X_n(t),$$

where c_i $(1 \le i \le n)$ are constants of the system taking values 0 and 1 and the arithmetic in (2) is modulo 2 addition. It is clear therefore that the system is completely specified by (a) the initial vector $x(0)$ and (b) the set of constants $c_1, \ldots, c_n$.

We also *assume* that $c_n \ne 0$; otherwise we could do without the register R_n.

The way the system works is that, on receiving a signal, each register does two things:

(i) It passes its contents along the row to its right-hand neighbour (the nth register, of course, cannot do this).

(ii) Those registers R_i for which $c_i = 1$ (which means they have leads to the mod 2 adder) also pass their content to the adder. This operates on them and passes the result to register R_1.

Once it is set up with the initial vector, the shift register can be regarded as the source of an infinite stream of binary digits

$$X_1(0), X_1(1), X_1(2), \ldots,$$

and so on, obtained by taking the successive contents of the first register. Although the stream produced is clearly not random, it can be shown to have what might naively be regarded as desirable properties of randomness, as listed below. Moreover it is mechanically easy to set up, and systems with over 200 registers are fast and compact generators of these pseudorandom sequences which can then be 'added' to the message bit-by-bit to simulate the action of a one time pad. Unfortunately, although widely used, this method will be shown to be extremely unsafe.

PROPERTIES OF SHIFT-REGISTER SEQUENCES

First we consider periodicity. An infinite sequence $(y_i : 0 \le i < \infty)$ is called *periodic* with *period* p if p is a positive integer such that $y_{i+p} = y_i$ for all i and, moreover, p is the smallest integer with this property. Thus, if the sequence $(y_i : 0 \le i < \infty)$ has period p, it can be written in the form

$$y_0, y_1, y_2, \ldots, y_{p-1}, y_0, y_1, \ldots, y_{p-1}, y_0, y_1, \ldots .$$

In other words, a sequence with period p is just a sequence of repetitions of a finite block of length p.

Returning now to the shift-register sequence: suppose that the initial vector $x(0)$ is not the zero vector and that the difference equations (1) and (2) are written in the form

(3)
$$x(t + 1) = Cx(t),$$

where C is the matrix

$$C = \begin{bmatrix} c_1 & c_2 & c_3 & \ldots & c_{n-1} & c_n \\ 1 & 0 & 0 & \ldots & 0 & 0 \\ 0 & 1 & 0 & \ldots & 0 & 0 \\ \vdots & \vdots & \vdots & & \vdots & \vdots \\ 0 & 0 & 0 & \ldots & 1 & 0 \end{bmatrix}.$$

Now, since we have assumed that $c_n = 1$, and since

$$\det C = c_n = 1,$$

we see that C is not singular. Iterating the relation (3) we get $x(t) = C^t x(0)$ and from this we can prove the following basic result.

Theorem 1 *The output sequence of a linear shift register is periodic and if there are n registers the maximum period is $2^n - 1$.*

Proof Since C is not singular, then so is C^i $(i = 0, 1, \ldots)$; also, $x(0)$ is non-zero and there are just $2^n - 1$ distinct non-zero n-vectors. So, if $m = 2^n - 1$, then

$$x(0), Cx(0), \ldots, C^m x(0)$$

are all non-zero and hence cannot all be distinct: say

$$C^s x(0) = C^{s+t} x(0),$$

where $0 \le s < s + t \le 2^n - 1$. But, since C is nonsingular, C^{-s} exists and so

$$x(t) = C^t x(0) = C^{-s} C^{s+t} x(0) = x(0).$$

Hence

$$x(r + t) = C^{r+t} x(0) = C^r C^t x(0) = C^r x(t) = C^r x(0) = x(r)$$

for all $r \ge 0$, and so C is periodic with period at most $t \le 2^n - 1$. Hence the output sequence is periodic. $\square$

Example   Consider the shift register of Fig. 2. If $x(0) = 1010$ then, as t runs from 0 to 15, we get

$$x(1) = 1101, \qquad x(2) = 0110, \qquad x(3) = 0011,$$

and so on until we arrive at

$$x(15) = 1010.$$

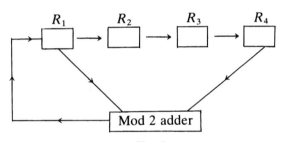

Fig. 2

In other words, we have a shift register of maximum period. $\square$

Not all feedback shift registers give sequences of maximum period, and those that do can be characterized by an elegant application of modern algebra.

We state without proof the main results.

Define the *characteristic polynomial* of the shift register to be the polynomial

$$P_n(x) = 1 + c_1 x + c_2 x^2 + \ldots + c_n x^n,$$

with $c_n \ne 0$, where the feedback coefficients c_i of the register are either 0 or 1. This characteristic polynomial is *primitive* if (a) it has no proper nontrivial factors and (b) $P_n(x)$ does not divide $x^d + 1$ for any $d < 2^n - 1$.

Note: Remember that we are working in the field of integers modulo 2, so that all polynomial operations are subject to this: e.g. $(x + 1)^2 = x^2 + 1$.

Theorem 2   *The output sequence of a linear shift register on a nonzero input has maximum period if and only if its characteristic polynomial is primitive.*

For a proof of this, see for example Lidl and Niederreiter (1986). Finding primitive polynomials is a nontrivial piece of modern algebra; it is enough to say that primitive polynomials exist for all n and have been extensively tabulated.

Exercises 8.2
1. Verify that, if the shift register having 4 registers and with feedback coefficients $c_1 = c_2 = c_3 = c_4 = 1$ is initiated with $x(0) = (0, 1, 1, 1)$, then $x(3) = (1, 1, 1, 0)$. What is its period?
2. Show that the characteristic polynomial of the shift register in Exercise 1 is not primitive.

8.3   The insecurity of linear shift-register sequences

As mentioned earlier, the output sequence of a linear shift register has sometimes been used as a source of pseudorandom numbers to take the place of a one-time pad. This is because they are easy to generate and also possess a certain superficial randomness as we now explain.

We define a *block* of *length* t to be a sequence of the form $011 \ldots 10$ containing exactly t ones. A *gap* of *length* t is a sequence of the form $100 \ldots 01$ with exactly t zeros.

We have the following result.

Theorem 1   *If a linear shift register with n registers has maximum period $2^n - 1$, then any output sequence of length $2^n - 1$ has the following properties:*
(a) *it contains exactly $2^{n-1} - 1$ zeros and 2^{n-1} ones;*
(b) *for any t, with $1 \le t \le n - 2$, it contains 2^{n-t-2} blocks of length t and the same number of gaps of length t.*

Proof   The state of the shift register at any time can be identified with an integer in the range 1 to 2^{n-1}: just take the ordered sequence of contents as the binary representation of the integer with the current output being the least significant digit.

Since all nonzero integers in the range 1 to $2^n - 1$ must occur as states in any maximum-length cycle, the result (a) follows from counting the number of even and odd integers in this set.

To prove (b), we note that a run of type $0111 \ldots 10$ with exactly t ones can occur as part of the output if and only if, at some stage in the computation, the state of the shift register is $0111 \ldots 10\, x_1 x_2 \ldots x_{n-t-2}$, where the x_i are 0 and 1. Since there are 2^{n-t-2} states of this form, and since each state will be realized at some stage in the computation because the register has maximum period, the result (b) follows for blocks. Interchanging 0 and 1 gives (b) for gaps. $\square$

Returning to the use of a sequence from a linear feedback register in encipherment, all that is usually done is the following. If

$$M = M_1 M_2 \ldots$$

is the message sequence of binary digits, and if $X_1 X_2 \ldots$ is the stream of digits produced by the shift register, then the cryptogram C is $C_1 C_2, \ldots$, where

(1)
$$C_i = M_i + X_i \pmod 2 \quad (1 \le i < \infty).$$

Thus, if M_i and C_i are known, X_i may be obtained trivially by

$$X_i = M_i + C_i \pmod 2.$$

Now consider a shift register with n registers and feedback coefficients $c_1, \ldots, c_n$. Once the enemy knows *any* $2n$ consecutive symbols X_i of the output stream, he can find these feedback coefficients by just solving n linear equations. This completely determines the cryptosystem, and hence we have proved the following result.

Theorem 2   *The use of output sequences from a linear feedback shift register as a pseudorandom one-time pad is insecure against known-plaintext attack.*

Exercises 8.3   1. In the proof of Theorem 2, show that the feedback coefficients $c_1, \ldots, c_n$ cannot necessarily be found if the known plaintext–ciphertext pair consists of $2n$ digits that are not consecutive.

2. Prove that, if $(X_1, X_2, \ldots)$ is a sequence of independent random variables with

$$P(X_i = 1) = p, \qquad P(X_i = 0) = q = 1 - p,$$

then the expected number of blocks of length t in the sequence $(X_1, \ldots, X_m)$ is $(m - t - 1)q^2 p^t$.
(Compare this with the property (b) of Theorem 1.)

8.4   Generating cyclic codes

Although linear shift registers are positively dangerous to use in a cryptosystem, they do have a very nice application in coding theory. Recall that a code $\mathscr{C}$ is *cyclic* if, whenever $\omega = (\omega_1, \ldots, \omega_n)$ is a codeword, then the shifted sequence $(\omega_n, \omega_1, \omega_2, \ldots, \omega_{n-1})$ is also a codeword.

Linear shift registers give a very attractive and easy way of generating cyclic codes because of the following theorem.

Theorem 1   *Suppose that the column vectors $x, Cx, C^2x, \ldots, C^kx = x$ are the successive iterations obtained from a linear shift register with n registers and a nonzero input x. If we take the $n \times k$ matrix*

$$H = \left[x, Cx, C^2x, \ldots, C^{k-1}x\right]$$

as the parity-check matrix of a binary linear code $\mathscr{C}$, then the code $\mathscr{C}$ is cyclic.

Proof   Suppose that ω is a nonzero codeword, so that

$$H\omega^T = 0.$$

Thus, if $\omega = (\omega_1, \ldots, \omega_k)$, we must have that

$$\omega_1 x + \omega_2 Cx + \ldots + \omega_k C^{k-1}x = 0.$$

Multiplying by C gives, since $C^k x = x$, that

$$\omega_k x + \omega_1 Cx + \ldots + \omega_{k-1} C^{k-1}x = 0,$$

which is equivalent to

$$H(\omega_k, \omega_1, \ldots, \omega_{k-1})^T = 0$$

and so $(\omega_k, \omega_1, \ldots, \omega_{k-1})$ is a codeword as required.   □

Exercises 8.4   1. Show that the three-register machine with characteristic polynomial $1 + x + x^2 + x^3$ generates a binary Hamming code.
2. What are the parameters of the cyclic binary code generated by the linear shift register whose characteristic polynomial is $1 + x^3$?

PROBLEMS 8

1. Prove that, for any source (no matter how correlated), if the message is encrypted using a one-time pad, then the symbols of the ciphertext are uniformly distributed and independent.

2. Consider the linear recurrence relation

$$s_{n+4} = s_{n+1} + s_n,$$

where $s_1 = 0$, $s_2 = 1$, $s_3 = 0$, and $s_4 = 1$, and all arithmetic is modulo 2.
   Construct a linear shift register, with suitable initial state $x(0)$, that will have $(s_n : n \geq 1)$ as its output sequence.

3. Prove that the pseudorandom number sequences constructed by the method of linear congruences as in Exercise 1.1 must be *ultimately periodic* in the sense that there exist integers p and N such that, for all $n \geq N$, $r_{n+p} = r_n$.

4. Prove that if a linear shift register has maximum period then this period may be achieved by any non-zero initial vector $x(0)$.

5. Consider a *linear recurring sequence* or *difference equation* of the form

$$s_{n+1} = a_1 s_n + a_2 s_{n-1} + \ldots + a_k s_{n-k} \pmod 2$$

where $a_i \in \{0, 1\}$ and the initial values $s_1 = b_1, \ldots, s_k = b_k$ are prescribed. Prove that the sequence $(s_{k+1}, s_{k+2}, \ldots)$ can be realized as the output sequence of a suitably chosen linear shift register. Can the sequence $(s_1, s_2, \ldots)$ always be realized?

6. Consider the linear recurrence sequences governed by

$$s_{n+5} = s_{n+1} + s_n \quad (n \geq 1).$$

Show that such sequences can be generated by a shift register having five registers and that there are initial vectors which result in sequences of periods 1, 3, 7, and 21, and that these are the only possible periods.

7. If $s_1 s_2 s_3 \ldots$ is the output sequence of a linear shift register with known coefficients $c_1, \ldots, c_n$, how many consecutive elements of the sequence (s_i) need to be known in order to determine the initial vector $x(0)$?

8. Show that, for suitable choices of initial non-zero vector, the four-register machine with coefficients

$$c_1 = c_3 = c_4 = 1, \qquad c_2 = 0$$

and the two-register machine with coefficients

$$c_1 = c_2 = 1$$

can produce the same output sequence.
   Can two distinct machines with the same number of registers and the same non-zero initial vector produce the same output sequences?

9. Let V_n denote the space of 0–1 n-vectors. Show that a given shift register with n registers induces an equivalence relation $\sim$ on V_n by: $u \sim v$ if the shift register, when started off with $x(0) = u$, achieves $x(t) = v$ for some non-negative integer t.

10. Given a linear shift register with n registers, show that, for any n-vector $(a_1, \ldots, a_n) \in \{0, 1\}^n$ and positive integer t, there exists an initial state $x(0)$ such that $x(t) = (a_1, \ldots, a_n)$.

11. The *general linear group* $GL(n, q)$ is the multiplicative group of $n \times n$ nonsingular matrices over a field of q elements. Prove that $GL(n, q)$ contains

$$q^{n(n-1)/2}(q-1)(q^2-1)\ldots(q^n-1)$$

distinct elements. Hence or otherwise show that, whatever the initial state, the period of a sequence from a linear shift register with n registers must divide

$$2^{n(n-1)/2}(2^2-1)(2^3-1)\ldots(2^n-1).$$

12. If C is the matrix of an n-register shift register and k is the order of C in $GL(n, 2)$, show that, for any initial vector $x(0)$, the output sequence has a period p which divides k.

13. Construct a primitive polynomial of degree 6 and hence a six-register machine with maximum period.

14. If a linear feedback shift register has output $(s_i : 0 \le i < \infty)$, show that

$$G(x) = \sum_{k=0}^{\infty} s_k x^k$$

is divisible by $(1 - x^p)^{-1}$, where p is the period of the output sequence.

9
Computational complexity

9.1 The intrinsic difficulty of a problem

Proving that there exist undecidable problems and noncomputable functions in mathematics is one of the most striking and important achievements of this era.

A cryptosystem whose decoding problem involved computing a noncomputable function would be in an impregnable position. However, it is easily seen that such a Utopia cannot exist: all the systems are finite and hence can, in principle, be cracked by exhaustive search of all possibilities.

The theory of computational complexity is concerned with the class of problems that can be solved in principle; but, within this class, the theory attempts to classify problems according to their computational difficulty measured as the amount of time or space their solution would take.

An understanding of the basic concepts of complexity theory is essential for cryptography, and in this chapter we will attempt to cover the main issues. First we proceed informally with some examples.

Example *Multiplication of integers.* Consider the problem of multiplying two *n*-bit binary integers x and y. If $x = x_1 \ldots x_n$ and $y = y_1 \ldots y_n$, then the standard method of 'long multiplication' which is taught at elementary school proceeds as follows:

Successively multiply x by y_1, y_2, and so on, shift, and then add the result.

Each multiplication of x by y_i takes about n single *bit operations*. Similarly the addition of the n products takes of the order of n^2 bit operations. Hence the total operation takes $O(n^2)$ bit operations.

Can we do better? That is, does there exist a faster algorithm in the sense of taking substantially fewer bit operations.

Well, a reasonably obvious approach is to think what you would do if confronted with two 500-digit numbers to multiply. Break them up into smaller numbers and then multiply the component parts. Put more precisely: given two n-bit integers, with n even, write

(1)
$$x = a2^{\frac{1}{2}n} + b, \qquad y = c2^{\frac{1}{2}n} + d;$$

then the product z can be formed by just three multiplications of $\frac{1}{2} n$-bit numbers by using the representation

(2)
$$z = xy = (ac)2^{n} + [ac + bd - (a - b)(c - d)]2^{\frac{1}{2}n} + bd.$$

Thus, if we let $T(n)$ denote the time it takes to multiply using this method, we have, since multiplying by 2^n is quick, that

(3)
$$T(n) \leq 3T(\tfrac{1}{2}n) + O(n).$$

solving this recurrence relation shows that

$$T(n) \leq An^{\log_2 3} + Bn,$$

where A and B are constants.

Since $\log_2 3 \simeq 1.59 \ldots$ this gives an algorithm of time complexity $O(n^{1.59})$ as compared with the standard $O(n^2)$ algorithm.

In fact the best algorithm known, due to Schönhage and Strassen (1971), has complexity $O(n \log n \, \log\log n)$ as compared with the obvious 'look at all the data' lower bound of $O(n)$. $\qquad \square$

Example 2 *Determinants and permanents.* Consider the following two computing problems. For any given input of an $n \times n$ matrix A, all of whose entries are 0 and 1, we wish to calculate:
(a) the determinant of A, denoted by $\det A$,
(b) the permanent of A, denoted by $\operatorname{per} A$ and defined by

$$\operatorname{per} A = \sum_{\pi} a_{1\pi(1)} a_{2\pi(2)} \cdots a_{n\pi(n)},$$

where the sum is over all permutations π of $(1, \ldots, n)$ and a_{ij} denotes the (i, j) entry of the matrix A.

On the surface, the permanent is a much simpler function of A than the determinant; it is the sum of the same collection of terms, without any of the hassle with $\pm$ signs as in the determinant's expansion.

However, from the computational viewpoint, the reverse is true: while the determinant is a relatively easy function to compute, the permanent has been shown to be almost certainly extraordinarily difficult. (The significance of the phrase 'almost certainly' should become apparent later.)

More precisely: the standard Gauss elimination method of computing the determinant of an $n \times n$ matrix involves $O(n^3)$ bit operations; while V. Strassen, in a fundamental paper in 1969, exhibited an algorithm which took

$$O(n^{\log_2 7}) = O(n^{2.81\cdots})$$

bit operations. Reducing the exponent below $\log_2 7$ has proved difficult, and the fastest current algorithm, due to Coppersmith and Winograd (1987), involves $O(n^{2.3976\cdots})$ operations. Since it is easy to see that all the entries in the matrix must be read, this implies

(4)
$$n^2 \le t_{\text{det}}(n) \le Cn^{2.3976\cdots}$$

where $t_{\text{det}}(n)$ denotes the time complexity of the problem of evaluating an $n \times n$ determinant and C is a constant.

For the permanent on the other hand, nothing like this is known. The fastest known algorithm is little better than checking all $n!$ possible terms in the term by term expansion.   □

Example 3 *Sorting.* Suppose we are asked to devise an algorithm which, when presented with n integers $a_1, \ldots, a_n$, will sort them into ascending order. An easy, almost instinctive approach is the following, known as *Bubblesort*.

Successively compare a_1 with each of $a_2, \ldots, a_n$. For maximum i: $a_1 > a_i$, place a_1 after a_i in the ordering. After $n - 1$ comparisons, this gives a new ordering $b_1, \ldots, b_n$. Repeat this procedure iteratively on this new ordering; it is easy to see that, after $n - 1$ such iterations, we have an ordered list. A simple count shows that this takes $O(n^2)$ comparisons.

However, there are several recursive algorithms based on the divide-and-conquer principle of sorting two halves of the set and then merging the sorted lists which take only $O(n \log n)$ comparisons. To get some idea of the difference, it is interesting to compare the values, as n varies, of these two functions:

n	$n \log_2 n$	n^2
50	~300	2500
500	~4500	250 000

Of course, the $O(n \log n)$ expression may 'hide' a large constant term, but nevertheless there is a striking difference, especially since sorting is a frequently used algorithm and the size of lists is often large.

One other striking point about sorting is that there is a lower bound of the same order on any comparison-based algorithm. This is often called the information-theoretic lower bound, but it is really just a direct consequence of the observation that any comparison-based algorithm can be represented by a binary tree-like structure and, in order to cover all possible $n!$ orderings, any such tree must have at least $n!$ leaves. □

Example 4

Primality testing. At first sight, testing whether an integer N(say) is a prime has a very quick easy algorithm: test whether N is divisible by 2 or by any odd integer in the range 3 to $N^{\frac{1}{2}}$. Since this only involves about $\frac{1}{2}N^{\frac{1}{2}}$ divisions, it is clearly (low) polynomial in N and hence fast.

However, further reflection shows that the representation of N to the computer would be by a binary string of length $\lceil \log_2 N \rceil$ and hence, in order that a prime-testing algorithm be regarded as fast, its complexity must be polynomial in $n = \lceil \log N \rceil$. So far, no algorithm meeting this requirement is known. However, over the last ten years or so, there have been some notable developments in the speed of such algorithms. There are now algorithms for testing whether N is prime which have a running time

$$t(N) = O(\ln N)^{c \ln \ln \ln N},$$

where c is a positive constant.

The exponent $c \ln \ln \ln N$ is an extremely slowly growing function, for example, if N has a million *decimal* digits, then $\ln \ln \ln(N)$ is only 2.68. Riesel (1985) reports an algorithm based on these methods, which, when implemented on the CDC Cyber 170–750 Computer is able to deal with 100-decimal-digit numbers in about 30 seconds and 200-digit numbers in $\simeq 8$ minutes. These modern algorithms are pretty complicated, and we will not attempt to describe them here. What is more practical and very fast is the idea of a *randomized* prime testing algorithm due to Rabin (1976, 1980) and Solovay and Strassen (1977). This we will describe in Section 6. □

Example 5

Greatest common divisors and the Euclidean algorithm. Consider the problem of finding the greatest common divisor (gcd) of two integers u and v. An obvious method is to factorize both integers into their prime components

$$u = 2^{u_1}3^{u_2}5^{u_3}\ldots, \qquad v = 2^{v_1}3^{v_2}5^{v_3}\ldots,$$

and then read off the gcd by

$$w = \gcd(u, v) = 2^{w_1}3^{w_2}\ldots$$

where $w_i = \min\{u_i, v_i\}$.

However, this is very inefficient. For a start, this involves factorizing both numbers and (as we shall see later) this cannot be done quickly for large integers. A method which avoids this is the familiar method known as *Euclid's algorithm*.

Assuming $u > v > 0$, this proceeds by a sequence of divisions:

$$u = a_1 v + b_1, \qquad 0 \le b_1 < v,$$
$$v = a_2 b_1 + b_2, \qquad 0 \le b_2 < b_1,$$
$$b_1 = a_3 b_2 + b_3, \qquad 0 \le b_3 < b_2,$$
$$\vdots \qquad\qquad \vdots \quad \vdots$$
$$b_{k-2} = a_k b_{k-1} + b_k, \qquad 0 \le b_k < b_{k-1},$$

until either $b_k = 0$ or $b_k = 1$. If $b_k = 1$, the integers u and v are coprime; if $b_k = 0$, then $\gcd(u, v)$ is b_{k-1}.

This algorithm is easily shown to be correct, and detailed analysis of its performance shows that the worst performance (measured by the number of divisions) of the algorithm occurs when u and v are successive *Fibonacci numbers* F_{n+2} and F_{n+1} respectively. In this case,

(5)
$$F_{k+2} = F_{k+1} + F_k,$$

and this leads to the following result of Lamé (1845) (see Knuth, 1969).

Theorem 1 *If $0 \le u, v < N$, then the number of division steps when Euclid's algorithm is applied to u and v is at most*

$$\lceil \log_\phi(\sqrt{5}\,N) \rceil - 2$$

where ϕ is the golden ratio $\frac{1}{2}(1 + \sqrt{5})$.

For a proof of these statements and a fascinating account of other gcd algorithms we refer to Knuth (1969).   □

Exercises 9.1

1. Prove that any algorithm which decides whether or not an $n \times n$ matrix is nonsingular has complexity at least n^2.
2. Estimate the complexity of the problem of finding which of m n-bit binary codewords has maximum weight.
3. Show that 2^n can be evaluated by $O(\log_2 n)$ integer multiplications.

9.2 P = polynomial time

A fundamental measure of the difficulty of a computation is the amount of time it takes. Formulating exactly what one means by 'time' is nontrivial; a rigorous formulation demands a very precise definition of the machine model, unit of time, and so on.

In the examples discussed in the previous section we measured the *complexity* of a computation in terms of the number of fundamental operations it took. These could be bit additions, comparisons, or whatever.

The key concepts are that:

(a) the complexity be regarded as a function of the input size (usually denoted by n),

(b) for a given input size n, the complexity is the time taken *in the worst possible* case.

Both these remarks need amplification. As far as (a) is concerned, consider the phrase 'input size'. When testing the primality of the integer N, we are obviously going to get a different complexity if we regard the input as of size N or as its more succinct representation as $n = \lceil \log_2 N \rceil$ binary digits. Because of this, *input size* is always regarded as the 'natural' length of an economic input.

As for defining the complexity somewhat pessimistically by its worst possible case, the other possibility—'average case'—is fraught with difficulties, both practical and theoretical. Not the least of these is the difficulty in deciding sensible prior distributons on the collection of inputs.

First we proceed very informally. An algorithm $\mathcal{A}$ is said to have *polynomial time complexity* if there exists some polynomial $p(x)$ such that

$$t_{\mathcal{A}}(n) \leq p(n),$$

for each integer n, where $t_{\mathcal{A}}(n)$ denotes the maximum 'time' taken by the algorithm over all inputs of size n.

A problem can be done in *polynomial time* if there exists some algorithm which solves the problem and which has polynomial time complexity; in this case, we say the problem *belongs to the class* P.

To illustrate the concept we look back at Examples 1–5 of the preceding section:

Example 1 Multiplication of integers is a polynomial-time operation.

Example 2   Evaluating a determinant is clearly in P. Evaluating a permanent is not known to be in P: it is widely thought not to be in P, and a proof either way would be of major importance.

Example 3   Sorting can be done in $O(n \log n)$ comparisons and, since a comparison can be done in polynomial time, sorting is clearly in P.

Example 4   Testing primality of an integer is not known to be in P. The recent results of Adleman, Pomerance, and Rumely (1983) and Cohen and Lenstra (1984) mean that 'it is very nearly in P'.

Example 5   Finding the gcd of two integers of size $\leq N$ and therefore of $\log N$ bits can, by Theorem 1.1, be done in time $O(\log N)$ and is therefore in P.

The class P is now fundamental in mathematics and computer science, and membership of it is usually regarded as meaning the problem is computationally tractable.

Although this is not strictly true (see Exercise 2 below), the following statements do show that it is an attractive and effective concept.

(1) The class P is robust with respect to different representations of the input, provided that such changes are polynomially related.†
For example, whether we regard the input of an $n \times n$ matrix as being size n or n^2 makes no odds.

(2) The class P is robust with respect to the machine model used. In other words, whether one is using a high speed computer, home micro, random-access machine, or even a Turing machine (described below), the class is unchanged. This is easy but messy to prove. It involves checking the time taken to simulate the basic operations of one machine on another.

For those who need a formal definition of P we now give a brief but rigorous description.

TURING MACHINES—A FORMAL DEFINITION OF P

A *Turing machine* has a 2-way infinite tape divided into squares as shown in Fig. 1. Each square can have a symbol from a finite *alphabet* Σ written on it. The tape is inspected or *scanned*, one square at a time, by a *read–write head*. The machine can be in one of a finite

---

† Two functions f and g are *polynomially related* if there exist polynomials p_1 and p_2 such that $f(n) \leq p_1(g(n))$ and $g(n) \leq p_2(f(n))$ for all sufficiently large n.

Tape

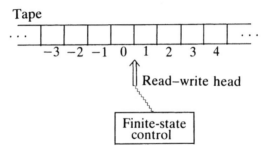

Fig. 1

set Q of *states*, $Q = \{q_0, q_1, \ldots, q_m\}$, and the machine's action at a given time is uniquely determined by its internal state and the symbol in the currently scanned square. The action will be to do any of the following operations:

(i) change the scanned symbol to another symbol from Σ,

(ii) move the read head one square to the right or left,

(iii) change its current state from q_i to q_j.

A *computation* on the machine consists of

(a) presenting it with a finite string $x \in \Sigma^*$ which is placed in the tape squares 1 to n (where n is the number of symbols in x).

(b) allowing the machine to start from its initial state (usually q_0) with its read–write head scanning square 1 and to proceed with its basic operations (read, write, change state) until it stops in its final state (q_f).

The *output* of the machine M on input x is the content of the tape when it has reached its final state. A *(single) step* in the computation consists of a single action (i)–(iii) above and the *length* or *time taken* by the computation is the number of such steps. If M denotes the particular Turing machine, we denote this time by $t_M(x)$.

A function $f : \Sigma^* \to \Sigma^*$ is *computable* by the Turing machine M, if, for all $x \in \Sigma^*$, whenever x is the input to M, it eventually halts with $f(x)$ on its output tape. Thus, a Turing machine is exactly analogous to a *programmed* computer; indeed, for any computation which can be carried out on a modern-day computer, there exists a Turing machine that will do exactly the same job.

In practice, constructing a Turing machine capable of even simple computations is pretty time-consuming (try building a Turing machine for multiplying two integers!), and so we develop a stock of basic machines which do tasks involved in a variety of computations. Then, in constructing a complex Turing machine, we use machines that have already been constructed, in the same way that subroutines are used in ordinary computer programmes.

We can now formally define time complexity. If M is a Turing machine which halts for all $x \in \Sigma^*$, the *time complexity* of M is the function $t_M : \mathbb{Z}^+ \to \mathbb{Z}^+$ given by

$$t_M(n) = \max\{t : \text{there exists } x \in \Sigma^* \text{ with } |x| = n$$
$$\text{and the time taken by } M \text{ on input}$$
$$x \text{ is } t\}.$$

A function f is computable in *polynomial time* or has *polynomial-time complexity* if there exists some Turing machine M that computes f, and some polynomial p such that $t_M(n) \leq p(n)$ for all n.

Now, it would be useless if a function were computable in polynomial time on one type of machine but not on a different model (e.g. a modern-day computer). It is pretty easy to see why this is not true. A basic comparison or bit operation is not going to blow up the time of a computation exponentially. Accordingly we may speak of a problem *belonging to* P as meaning that there exists for the problem some algorithm working in polynomial time measured in terms of basic operations such as adding, multiplying, comparing, etc. Provided that each of these basic operations can be done in polynomial time, the overall algorithm will work in polynomial time. Thus, in practice, to verify that an algorithm is in P, we rarely consider the Turing model but work at a much higher level.

Exercises 9.2
1. Prove that the problem of solving a set of linear equations belongs to P.
2. A *clique* of size k in a graph G is a collection of k vertices that are all joined to each other. Show that for any fixed k there is a polynomial algorithm for finding such a clique. (*Note*: Thus, for any fixed k, say $k = 1000$, the problem is in P. However, there is no known practical algorithm for this problem for such large values of k. This illustrates the remark that membership of P does not guarantee feasibility.)

9.3 NP = nondeterministic polynomial time

In contrast to P, which is relatively easy to comprehend, the class NP is a much more complex object. As with P we will first describe the basic idea informally and then define it rigorously.

Suppose you are in the business of selling large composite numbers to very busy purchasers. With the aid of slaves working unlimited hours beforehand you can build up a stock of these composite numbers $c_1, c_2, \ldots$ and, in order to be able to sell them quickly, you will also have a corresponding stock of factors $y_1, y_2, \ldots$ so that, when it comes to a sale, all you have to do is give the number c_i together with its factor y_i. All that has to be done to verify that c_i is composite is an elementary division by y_i. In other words, compositeness is quickly verifiable. The y_i is called a *certificate* of the nonprimeness of c_i because, by using it, there is a polynomial-time algorithm for checking that c_i is composite.

Informally, a property is said to *belong to* NP if possession of that property can be checked in polynomial time with the help of an appropriate certificate.

Example A *graph G* on *n vertices* is a collection of *n* points and *edges* joining some of these vertices. The graph *G* is said to be *Hamiltonian* if there exists a simple circuit passing once and only once through each of these points. For example, in Fig. 1, G_1 has the Hamiltonian circuit (a, b, d, e, g) but the graph G_2 has no Hamiltonian circuit.

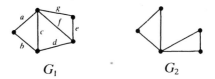

G_1 G_2

Fig. 1

Testing whether a graph has or has not a Hamiltonian circuit is a difficult computational problem. No polynomial-time algorithm is known. However we can say:

(a) The property of being Hamiltonian is in NP.

Proof Use the circuit as a certificate.

(b) The property of being non-Hamiltonian is not known (or indeed thought) to be in NP.

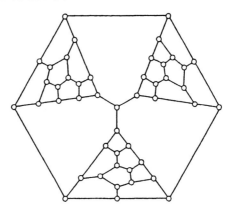

Fig. 2

To add authenticity to (b), try convincing someone that the graph of Fig. 2 is not Hamiltonian, and then estimate how long it would take you to 'sell non-Hamiltonian graphs' with (say) 1000 vertices. ☐

This example illustrates a fundamental difference between our states of knowledge about P and NP respectively. In the case of P, a property belongs to it if and only if its negation does. Whether or not this holds for NP is a major unsolved problem in the theory of computer science.

Having hopefully got across the idea of NP, we define it formally as follows.

First we define a *property* or *language* π as any subset of Σ^* (the collection of finite strings from the alphabet Σ). We then say that $\pi \in \text{NP}$ if there exists a function $f : \Sigma^* \times \Sigma^* \to \{0, 1\}$ such that

(a) $x \in \pi$, if and only if there exists $y \in \Sigma^*$ such that $f(x, y) = 1$,

(b) f is computable in time bounded by a polynomial in x.

In terms of Turing machines, what we are thinking of is using the 'y' associated with the input x as the *certificate* of membership by x of π, and letting the Turing machine operate in polynomial time on an input consisting of the genuine input x, together with the certificate y, as shown in Fig. 3.

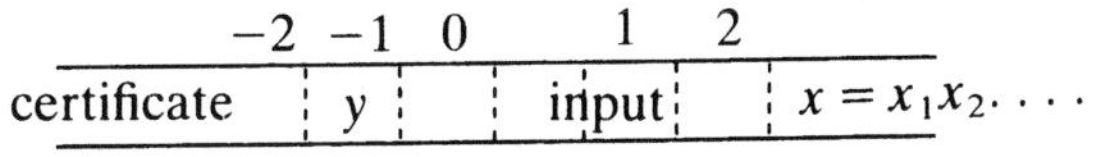

$$\begin{array}{c|ccccc} & -2 & -1 & 0 & 1 & 2 \\ \hline \text{certificate} & y & & \text{input} & x = x_1 x_2 \cdots \end{array}$$

Fig. 3

A trivial consequence of the definition is the containment

(1) $$P \subseteq NP,$$

provided that we regard P as a collection of properties. The question whether or not $P = NP$ is probably the most important and challenging problem in theoretical computer science; despite a vast amount of effort, there has been little real progress on this problem since it was first posed (circa 1970).

Exercises 9.3

1. Which of the following properties of an $n \times n$ matrix of zeros and ones can you show to be in NP:
   (a) having determinant at least n,
   (b) having permanent at least n,
   (c) having permanent at least $\frac{1}{2}n!$?

2. Which of the following properties of graphs can you show to be in NP:
   (a) having a path which uses each edge exactly once,

(b) having a path of prescribed length,
(c) having its maximum-length path a prescribed length?

9.4 NP-complete/hard problems

A crucially important property of NP is that it contains 'hardest properties' such that, if any one of these properties could be decided in polynomial time, then so could every property in NP.

To make this more precise, we say that a property π_1 is *polynomially reducible* to π_2 if there exists some function f which is in P and which has the property that x has property π_1 if and only if $f(x)$ has property π_2. We write this as

(1)
$$\pi_1 \propto \pi_2.$$

It is easy to see that:

(2) If $\pi_1 \propto \pi_2$, then the existence of a polynomial-time algorithm for π_2 implies the existence of a polynomial-time algorithm for π_1.

Proof Suppose $\mathscr{A}$ is an algorithm for π_2 which works in time $t(n)$. If x is any input to π_1, with $|x| = n$ (say), then transform x to $f(x)$ and apply $\mathscr{A}$. Since the transformation works in polynomial time, $|f(x)|$ is bounded in size by some polynomial $g(n)$. Thus the transformation and $\mathscr{A}$ together work in time bounded by some polynomial. □

We can now state the very important result of Cook (1971).

Theorem 1 *There exists a property π in NP such that any other property π' in NP is polynomially reducible to π.*

Any such π is called NP-*complete*.

In fact Cook proved more than this, and his seminal paper, followed shortly afterwards by that of Karp (1972), has led to the creation of a huge catalogue of NP-complete properties. See for example Garey and Johnson (1979), which contains over 3000 such properties drawn from fields as diverse as number theory, logic, combinatorics, and automata theory.

Indeed it is a curiosity that almost every well-known property in NP but not known to be in P seems to be NP-complete, though there

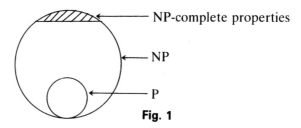

Fig. 1

is a theorem by Ladner (1975) to the effect that if $NP \neq P$ then $NP \setminus P$ contains properties which are not NP complete.

NP-HARD PROBLEMS

Consider the following problem:

(π) Construct an algorithm which, given a graph G, *counts* the number of Hamiltonian circuits of G.

Trivially, π is 'harder' than the pure existence problem of deciding whether or not G has a Hamiltonian circuit.

Problems such as this are called NP-hard. More formally, we define a problem π to be NP-*hard* if there exists some NP-complete property π_0 such that, if there exists a polynomial algorithm for π, then there exists a polynomial algorithm for π_0.

Trivial consequences of this definition are the following statements.

(1) If there exists a polynomial-time algorithm for any NP-hard problem, then $NP = P$.
(2) If π_1 is NP-hard and $\pi_1 \propto \pi_2$, then π_2 is NP-hard.
(3) Any NP-complete property is NP-hard.

The converse of (3) is not true. To understand this, we relate the classes P and NP with the class EXPTIME: the class of those problems that can be solved by an algorithm running in time $c^{p(n)}$ on inputs of size n, for some positive constant c and polynomial p.

It is trivial to prove the containment

$$P \subseteq NP \subseteq EXPTIME,$$

and it is known that

$$P \neq EXPTIME.$$

The sort of problems that are solvable in EXPTIME but not thought to be in NP—and hence *a fortiori* not in P—are of the following type.

Take a game such as Go or Chess or Draughts (= Checkers) and extend it in the natural way to a game on an $n \times n$ board (the extension of chess is not so natural). Deciding whether or not the first player has a winning strategy can be done in EXPTIME by just trying all possibilities.

1. A *clique* in a graph is a set of vertices that are all joined to each other; an *independent* set is a set of vertices no two of which are joined to each other by an edge. CLIQUE is the problem of deciding whether a graph has a clique of prescribed size, INDEPENDENT SET is the problem of deciding whether a graph has an independent set of prescribed size. Prove that

CLIQUE ∝ INDEPENDENT SET,   INDEPENDENT SET ∝ CLIQUE.

2. A *Hamiltonian path* of a graph is a path that visits every vertex exactly once. Assuming that the problem of deciding whether a graph has a Hamiltonian path is NP-complete show that the problem LONGPATH—deciding whether a graph has a path of prescribed length—is NP-hard. Is LONGPATH NP-complete?

3. A much studied NP-complete problem is SATISFIABILITY which is the problem of deciding whether there is a truth assignment for variables $x_1, \ldots, x_n$ that satisfies a given collection of clauses in these variables. For example, the collection of clauses

$$(x_1 \vee x_2 \vee \neg x_3) \wedge (\neg x_1 \vee \neg x_2 \vee \neg x_3) \wedge (x_1 \vee x_3)$$

is satisfied by setting $x_1 = x_2 = $ TRUE and $x_3 = $ FALSE. Show that

SATISFIABILITY ∝ CLIQUE.

4. Prove that, if problems $\pi_1, \pi_2,$ and π_3 satisfy $\pi_1 \propto \pi_2$ and $\pi_2 \propto \pi_3$, then $\pi_1 \propto \pi_3$.

9.5 Circuit complexity

We now turn to a completely different model of computation which corresponds roughly to the 'chip' or VLSI circuit. A *switching* or *combinatorial* or *logic circuit* is a device for computing Boolean functions. It is usually represented as a finite directed acyclic graph with a partition of its vertices into *input* vertices, *internal* vertices or *gates*, and one *output* vertex. Each input vertex is associated with one of the arguments with which the circuit is to compute. Each of the internal vertices or gates has an operation associated with it, such as ∨, ∧, or any other of the 16 Boolean functions of 2 variables. The output vertex is associated with the value of the function being computed.

Example   Fig. 1 shows the acyclic graph representation of a circuit computing

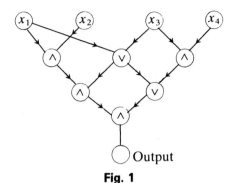

Fig. 1

the function

(1)
$$f_0(x_1, x_2, x_3, x_4) = [(x_1 \wedge x_2) \wedge (x_1 \vee x_3)]$$
$$\wedge [(x_1 \vee x_3) \vee (x_3 \wedge x_4)]. \qquad \square$$

A circuit C with n input vertices *computes* the function $f(x_1, \ldots, x_n)$ if, whenever the input gates are assigned $x_1, \ldots, x_n$, and these values are processed by the internal gates in the obvious order induced by the acyclic ordering of C, then the value at the output gate is $f(x_1, \ldots, x_n)$.

The *size* of the circuit C is its number of internal gates and is denoted by $c(C)$.

If f is any function, then the *circuit complexity* of f is defined to be the minimum size of a circuit that computes f and is denoted by $c(f)$. Thus, for f_0 defined by (1) in the example above, we know that

$$c(f_0) \le 6.$$

There is an obvious extension of this notion to general properties. We illustrate by example.

Example   (Hamiltonian circuit revisited.) Consider the question of deciding whether or not a graph G has a Hamiltonian circuit. For each value of n, it is possible to find a circuit C_n having $O(n^2)$ input gates (one for each possible edge) that gives output TRUE if and only if the input graph G has a Hamiltonian circuit. However, all such C_n known at present have an exponential number of gates. $\square$

Thus we say that a property π has *polynomial-size circuits* or has *small circuits* if there is a sequence of circuits $(C_n : 1 \le n < \infty)$ and a polynomial p such that C_n decides π on all possible inputs of size n and the size $c(C_n)$ satisfies

(2)
$$c(C_n) \le p(n) \quad (1 \le n < \infty).$$

It is important here to emphasize the relationship between circuit complexity and Turing complexity, which we sum up in the following statements:

(3)   If a problem is (Turing-)computable in polynomial time, then it has small circuits.

(4)   The converse is not true: there are nonrecursive functions that are not computable by *any* Turing machine but yet have small circuits.

(5)   If *any* NP-hard problem could be shown to have small circuits, then every problem in NP would have small circuits.

As far as effective computability and its cryptographic implications are concerned, any problem with small circuits would be regarded as 'easy' and not a secure basis on which to base a cryptographic system.

Exercises 9.5

1. A *Boolean function* of n variables is a map $\{0, 1\}^n \to \{0, 1\}$. If B_n denotes the class of such functions, show that

$$|B_n| = 2^{2^n}.$$

2. List the 16 Boolean functions of two variables and show that they can all be expressed in terms of combinations of the connectives

$$\wedge \text{ (and)}, \quad \vee \text{ (or)}, \quad \neg \text{ (not)}.$$

(Such a set of functions is called a *complete* set.)

9.6 Randomized algorithms

An algorithm which has an error probability of less than 2^{-100} and which, within a minute, identifies $2^{400} - 593$ as the largest prime below 2^{400} has an immediate practical and aesthetic appeal. Such is the primality testing algorithm of Rabin (1976), which, together with a similar one of Solovay and Strassen (1977), is the best-known example of a randomized algorithm. Its startling success has had a major impact on the theory of algorithms.

First we give an example to illustrate the basic idea. Suppose we have a polynomial expression in n variables, say $f(x_1, \ldots, x_n)$, and we wish to check whether or not f is identically zero. To do this analytically could be a horrendous computation. Suppose instead we generate a random n-vector $(r_1, \ldots, r_n)$ and evaluate $f(r_1, \ldots, r_n)$.

If $f(r_1, \ldots, r_n) \neq 0$, we know $f \neq 0$; if $f(r_1, \ldots, r_n) = 0$, then either f is identically zero or we have been extremely lucky in our choice of $(r_1, \ldots, r_n)$. Do this several times and, if we keep on getting $f = 0$, we conclude that f is identically zero. The probability of us having made an error is negligible.

This example illustrates the basic idea behind randomized algorithms. They are not to be confused with probabilistic approximation algorithms; they tend to be much more powerful, and correspondingly much rarer. Indeed, as we shall see, the existence of a randomized algorithm for a problem seems to correspond, in more than one sense, with the problem being tractable. For, given a particular problem, suppose that we can find a fast algorithm which is correct with probability $>1 - 2^{-k}$, where k is an integer parameter that we may choose independently of the size of the problem; then, by making k large enough, we can reduce the probability of error far below the probability of a hardware failure causing an incorrect answer (Adleman and Manders, 1977).

We shall give a more precise definition of terms such as 'randomly decidable' and 'randomized algorithm' later. For the moment, it is enough to proceed more loosely so that the flavour of such an algorithm can be assimilated by means of different examples.

If π is a computational problem and x is the input or instance of π under consideration, a *randomized algorithm* for solving π proceeds as follows. At certain junctures in the execution of the program for solving the instance x of π, the algorithm makes a random decision. With the exception of these random choices the algorithm proceeds purely deterministically.

For some instances, a randomized algorithm may sometimes produce an incorrect solution. We insist that any randomized algorithm has the property that the probability of error can be made arbitrarily small.

Example 1   *Checking polynomial identities.* Suppose that we are given a purported polynomial identity of the form $Q(x_1, \ldots, x_n) \equiv 0$, where, for simplicity, we assume the coefficients in Q are integers. For example, if we did not know Vandermonde's identity, we might need to check the truth of

$$Q \equiv \det \begin{bmatrix} 1 & x_1 & \ldots & x_1^{n-1} \\ \vdots & & & \vdots \\ 1 & x_n & \ldots & x_n^{n-1} \end{bmatrix} - \sum_{i<j} (x_i - x_j) = 0$$

for some fixed value of n, say $n = 100$. In this case, Q is a polynomial of total degree approximately 5000, and thus the deterministic method of simplifying by direct expansion is hopeless. The randomized algorithm which can be used is very elementary. It makes use of the following lemma.

Lemma   *Let $Q = Q(x_1, \ldots, x_n)$ be a polynomial in the variables $x_1, \ldots, x_n$ and suppose that Q is not identically zero. If I is any set of elements in the field of coefficients of Q such that $|I| \geq c \deg Q$, then the number of elements of $I \times \ldots \times I$ that are zeros of Q is at most $|I|^n/c$.*

Hence the following randomized algorithm enables us to check an identity of the form $Q \equiv 0$.
(1) Chooose I so that $|I| > c \deg Q$ with $c > 2$.
(2) Choose an integer N such that c^{-N} is small enough for our purposes (say N such that $c^{-N} < 2^{-400}$).
(3) Select N variables $(y_1, \ldots, y_n)$ at random from $I \times \ldots \times I$. If any one of these y is not a zero of Q, then $Q \neq 0$. If all are zeros of Q, then, by the above lemma, the probability that we make an error in asserting that $Q \equiv 0$ is very small. $\square$

Example 2   *Primality testing.* Solovay and Strassen (1977) and Rabin (1976) give different randomized algorithms which test a given integer n for primality. In both algorithms, the basic idea is the same and consists of taking k integers i uniformly distributed between 1 and $n - 1$, and then checking for each i whether a predicate $W(i, n)$ holds.

The exact form of $W(i, n)$ differs in the two papers but, in both cases, $W(i, n)$ has the properties:
(a) If n is prime, then $W(i, n)$ is false for all i.
(b) If n is composite, then $W(i, n)$ is true for some fixed positive fraction of the integers i in the range chosen.
(c) $W(i, n)$ can be tested quickly: it has a complexity bound growing slowly with the input size n. In this case, the bound is $\log n$.

The difficulty is in finding predicates $W(i, n)$ satisfying these conditions. The test devised by Rabin works as follows. Write $n = 1 + 2^t m$, where m is odd. For $1 \leq x \leq n$, write

$$x_0 = x^m \pmod{n}, \qquad x_j = x_{j-1}^2 \pmod{n} \quad (1 \leq j \leq t).$$

Thus

$$x_t = x^{n-1} \pmod{n}.$$

Theorem 1   *Let n be an odd composite integer greater than 4. Then for at least*

$\frac{3}{4}(n-1)$ of the x $(1 \leq x < n)$, either (i) $x_t \neq 1$ or (ii) for some j, with $1 \leq j \leq t$, we have $x_j = 1$ and $x_{j-1} \neq n-1$.

If n is prime, then neither (i) nor (ii) holds for any x $(1 \leq x < n)$.

Thus, in Rabin's method, the predicate $W(x, n)$ is taken to be that either (i) or (ii) holds. When a pair (x, n) satisfying (i) or (ii) is found, we call x a *witness* to the compositeness of n.   □

RANDOM POLYNOMIAL TIME

Motivated by the success of the Rabin–Solovay–Strassen algorithm for testing primality, it is natural to study the class of properties that are randomly decidable. Recall that we regard a *property* as a collection of strings from some input alphabet. Such a collection is termed a *language* L and we represent an input x having the given property by the statement $x \in L$.

A language L is *randomly decidable* in polynomial time or is in the *random-polynomial-time* class RP if there exists a polynomial f and a polynomial-time algorithm that computes, for each input x and each possible certificate y of length $f(|x|)$, a value $v(x, y) \in \{0, 1\}$ such that

(i) If $x \notin L$, then $v(x, y) = 0$ for all y.
(ii) If $x \in L$, then $v(x, y) = 1$ for at least half of all possible certificates y.

There are various points to note. First, since the class NP could be defined by replacing 'at least half' in (ii) by 'at least one' we know

$$RP \subseteq NP,$$

and obviously

$$P \subseteq RP.$$

Finally we remark that 'half' in (ii) can be replaced by α, where α is any fixed fraction with $0 < \alpha < 1$.

The importance of showing that a language L belongs to RP is that we can then recognize any string x as a member of L by the following procedure: (a) generate a random sequence y of length $f(|x|)$; (b) calculate $v(x, y)$; and (c) repeat (a) and (b) k times and return the answer '$x \in L$' iff $v(x, y) = 1$ for at least one of the k random y generated. Then, exactly as in the special case of the language of primes, the probability of error can be made arbitrarily small.

Thus, once we have shown that a language belongs to RP, we can essentially recognize it; unfortunately it seems exceptionally difficult to find randomized algorithms for problems for which no polynomial algorithm is known. The reasons for this are twofold. First, it is easy to prove the following.

(4)   RP = NP if and only if some NP-complete language is randomly decidable in polynomial time.

Hence, since it does seem unlikely that RP = NP, we are most unlikely to find a randomized algorithm for any NP-complete problem. Moreover, as we mentioned earlier, there seem to be few 'natural' problems that are thought to be in $NP \setminus P$ and that have not been shown to be NP-complete.

We close this section by proving a result of Adleman (1978) which gives added justification to our claim that membership of RP corresponds to practical tractability.

Theorem 2   *If a language is randomly decidable, then it has polynomial-size circuits.*

The proof of this intriguing result is by a refreshingly simple counting argument.

Proof (Sketch)   If the language L is in RP, then it is easy to see there exists a polynomial t such that there can be at most $2^{t(n)}$ possible witnesses (certificates) to membership in L of any input x of size n that belongs to L.

Take $m = 2^{t(n)}$ and let $(w_j : 1 \leq j \leq m)$ be these possible witnesses. Form the incidence matrix $[a_{ij}]$ in which the rows correspond to the possible inputs and a_{ij} is 1 or 0 depending on whether or not w_j is a witness to the membership of input x_i in L.

Because $L \in$ RP, it follows, by definition of RP, that at least half the w_j are witnesses to a given x, and this means that in each row of this matrix, at least half the entries are 1. Hence at least half the entries in some column must be 1. Without loss of generality, let this column correspond to w_{i_1}. Remove it and all rows x_i for which it is a witness to membership of x_i in L. Call the reduced matrix $A^{(1)}$, and note it must have exactly the same property. Hence there exists another witness w_{i_2} which will witness membership of at least $\frac{1}{2}$ the remaining x_i and so on. Continuing to iterate this procedure, we see—by a simple counting argument—that, after at most a polynomial number of iterations, we have found a polynomial number of witnesses $w_{i_1}, w_{i_2}, \ldots$ which together witness membership of L for each of the x_i. These can be used to build a polynomial size circuit for membership of L thus proving the theorem. $\qquad\square$

Note The nonconstructive nature of the proof, given that a problem is in RP, gives you no idea how to build these polynomial-size circuits.

Exercises 1. Show that, if Rabin's randomized algorithm for primality testing is applied
9.6 to a composite integer N, then the expected number of random integers tried before the algorithm finds a witness that declares N composite is less than or equal to $\frac{4}{3}$.

2. Show that, if $\pi_1 \propto \pi_2$ and if π_2 can be solved by a randomized algorithm in polynomial time, then π_1 can be solved by a randomized polynomial algorithm.

9.7 Effective versus intractable computations

It is appropriate now to take stock of the situation as outlined in the previous sections.

Having started off with the notion of membership of P as corresponding to tractable, we have now widened this class to include those problems having polynomial-size circuits and hence, by Adleman's theorem, those having randomized polynomial-time algorithms.

As far as cryptography is concerned, we have to regard a problem having small circuits as being tractable. If C denotes the class of problems having small circuits, the interrelationship among the various classes is believed to be as illustrated in Fig. 1. All that is definitely known to be true is the following: the containments as illustrated hold, the class P is a proper subset of EXPTIME, and C is not a subset of EXPTIME.

Remark The problems in C\EXPTIME are based on nonrecursive sets and, as far as we know, are of no practical significance at this level.

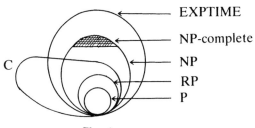

EXPTIME

NP-complete

NP

RP

P

C

Fig. 1

Henceforth we shall regard a problem as *tractable* if it has small circuits and regard an algorithm as being *effective* if it can be realized by a circuit with only a polynomial number of gates.

Note: This is a different use of 'effective' from its use in recursive function theory.

The cryptographic ideal would be to have a system such that the encrypting/decrypting functions were effective, but the cryptanalyst without the key would be faced with a computationally hopeless problem. In this way, computational complexity is a formalization of Shannon's notion of the 'work characteristic' of a cryptosystem.

The relationship between NP and P, together with a huge catalogue of NP-complete problems and the interrelationship between the various complexity classes, is beautifully presented in the monograph of Garey and Johnson (1979). For a wider discussion of complexity issues we refer to Aho, Hopcroft, and Ullman (1974) or, at a more advanced level, Wagner and Wechsung (1986).

PROBLEMS 9

1. What is the complexity of the Huffman encoding algorithm measured as a function of n, the number of sourcewords to be encoded?

2. Consider the problem of finding $A_2(n, d)$: the maximum size of a set of binary codewords of length n and minimum distance d. Estimate the complexity of the brute-force algorithm of finding $A_2(n, d)$.

3. A communication network can be regarded as a set of n points (stations), some of which are joined by a direct link (edge). Devise an algorithm of complexity $O(n^2)$ which will find the route containing fewest edges linking any two prescribed stations.

4. Consider the problem of finding both the maximum *and* the minimum of a set of n elements using only pairwise comparisons. The trivial method uses about $2n$ comparisons. Use the 'divide and conquer' principle to devise an algorithm that uses about $\frac{3}{2}n$ comparisons. Prove that your algorithm achieves this bound for the case when $n = 2^r$.

5. The *covering number* β of a graph G is the minimum number of vertices in a set U such that each edge of G has at least one endpoint in U. Prove that determining β is NP-hard.
   [Hint: Reduce the problem to that of finding a clique of given size.]

6. Prove that if a property π is in NP, then there exists a polynomial p such that, for any input of size n, deciding whether or not it has π takes time at most $2^{p(n)}$.

7. Construct a circuit, with a minimum number of gates, that computes the function

$$f(x_1, x_2, x_3, x_4) = (x_1 \vee x_2) \vee (x_2 \wedge x_3) \vee (x_1 \wedge x_4).$$

8. Suppose that $f(x_1, \ldots, x_n)$ and $g(x_1, \ldots, x_n)$ are Boolean functions in the variables $x_1, \ldots, x_n$. If h is the function defined by

$$h(x_1, \ldots, x_n) = f(x_1, \ldots, x_n) \vee g(x_1, \ldots, x_n),$$

show that

$$C(h) \leq C(f) + C(g) + 1.$$

Give an example to show that strict inequality can hold in this.

9. The *depth* of a Boolean circuit is the maximum number of gates in a path from an input to the output. Show that if a circuit has depth d and size c than

$$c \leq 2^d - 1.$$

10. Consider the Boolean function f defined for all even $n = 2m$ by

$$F(x_1, \ldots, x_m, y_1, \ldots, y_m) = \begin{cases} 1 & \text{if } x_i = y_i \ (1 \leq i \leq m), \\ 0 & \text{otherwise.} \end{cases}$$

Show that the circuit complexity of f is $n - 1$.

11. The problem of finding the minimum distance of a linear code has been formulated in Chapter 3. In terms of the parity-check matrix, it reduces to: given a matrix H and positive integer w, is there a nonzero binary vector x, with $|x| < w$, such that $Hx^T = 0 \pmod 2$. Prove that this is an NP-complete problem.

[Berlekamp, McEliece, and van Tilborg (1978)]

12. Show that testing whether an integer i, with $1 \leq i < n$, is a witness to the compositeness of n, using Rabin's test, has complexity which is a polynomial function of $\log n$.

13. Show, by computer search, that the fraction of 'Rabin witnesses' to the compositeness of 200043671 is 0.7507.

(This shows that the bound $\frac{3}{4}$ in Theorem 6.1 is tight.)

(Rabin, 1980)

10

One-way functions

10.1 Informal approach: the password problem

The idea underlying the concept of a one-way function is at the heart of cryptography. However, it is very difficult to give it a precise mathematical definition. Informally, a *one-way function* is a function $f: S \to T$, where S and T are any two sets, such that:

(1) for any $x \in S$, $f(x)$ is 'easy' to compute,

(2) given the information that $f(x) = y$, there is no 'feasible' way of obtaining (computing) x for a reasonably large proportion of the y belonging to T.

The operative words here are 'easy', 'feasible', and 'reasonably large'.

It is clear that, given $f(x)$, one way of obtaining x is to search exhaustively through all possible values of $x \in S$. We do not regard this as feasible, since typically S will be the collection of all n-bit binary strings with $n \sim 200$.

We require essentially that the computation required to find x from y be too time-consuming or costly to be practical, whenever y belongs to some 'fairly large' subset of T.

Example   An elementary example (Purdy 1974) of a candidate one-way function is to take any large prime p, and to let $f(x)$ be any polynomial over the field of integers mod p. Then it is relatively easy to calculate $f(x)$ $(1 \le x \le p - 1)$, but usually hard to find a solution of

$$f(x) = y. \qquad \square$$

The imprecise definition as given above means that what is a one-way function varies with time. For example, as Pohlig and Hellman (1978) point out, a computation requiring one million instructions and 10 000 words of memory would not have been regarded as easy in 1950, but in 1978 could have been carried out in a few seconds on a small computer. Thus a function that might have been regarded as one-way in 1950 would certainly not be one-way today.

One method of giving a formal definition might to be take a physical approach. For example, as claimed by Keyes (1975), a 10^{60}-bit memory will always be unattainable because, even if only one molecule is needed per bit of memory, its construction would require more mass than exists in the solar system. In the same vein, thermodynamics places a limit of 10^{70} on the number of operations that can be performed, even if the entire energy of the sun could be harnessed forever.

Demanding that a computation be regarded as infeasible only if it contravened these limits would rule out any of the functions which we would like to use in practice. A lower level of stringency is to use the ideas of computational complexity introduced in the last chapter. First let us consider some of the properties we would like a one-way function f to possess.

(I) Computing $f(x)$ from x must be feasible: we express this by demanding that f is computable in polynomial time, that is, f belongs to P.

(II) Computing f^{-1} must not be easy; hence we stipulate that no polynomial time algorithm for computing f^{-1} is known.

(III) As a third condition, we demand that the function f be *honest*: this means that, for some polynomial p, $|f(x)|$ is guaranteed not to be so small that $|x| > p(|f(x)|)$.

This last condition of honesty is a somewhat technical condition to eliminate functions such as $f(x) = \lceil \log \log x \rceil$, which would certainly satisfy (I) and (II), since even writing down the answer of the inversion would take exponential time, but which would not normally be regarded as useful one-way functions.

A function f satisfying (I), (II), and (III) we call a *weak one-way* function.† The reasons for wanting even more will be evident from the next two examples.

Example 1   Let I_k denote the set of all k-bit integers, that is, $I_k = \{2^{k-1}, \ldots, 2^k - 1\}$ ($k = 1, 2, \ldots$). Let $S_k = I_k \times I_k$ and let $f : S_k \rightarrow \mathbb{Z}^+$ be defined by

$$f(m, n) = mn.$$

If we let $S = \bigcup \{S_k : 1 \leq k < \infty\}$ and define f on S by its obvious extension, then we have what is commonly regarded as a weak

---

† These conditions (I–III), together with the requirement that f be 1–1, are what Grollman and Selman (1985) call 1-way. However this is really not enough for cryptographic purposes.

one-way function. At the present time, inverting f, which is clearly just the problem of factoring, is certainly not known to be in P. ☐

Note When we are speaking of polynomial time in this sort of problem where the input 'x' is an integer which will be represented by $\lceil \log x \rceil$ bits, we demand that the running time of any polynomial algorithm must be bounded by a polynomial in $\lceil \log x \rceil$.

THE PASSWORD PROBLEM

One of the first applications of one-way functions was to solving the problem of the security of passwords in an electronic authentication scheme.

Consider, for example, a remotely accessed computer system in which each user must log on by identifying himself to the system by sending a secret password. It is dangerously insecure to store a list of users and their passwords in an undoctored form on a file in the machine, since it would be almost impossible to keep the file contents secret. R. M. Needham (see Wilkes, 1972, p. 91) proposed the following easy solution to this problem.

Let f be any one-way function whose domain is the set of possible passwords (≡strings of alphabetic symbols).

When a new user U_i joins a system, it is customary for him to choose a password, say P_i. Instead of the computer storing P_i, it stores the easily computed string $f(P_i)$.

Any intruder is now welcome to the file containing the list of user names $((U_i, f(P_i)) : 1 \leq i \leq N)$. There is no need to keep it secret. When user U_i returns to reuse the machine, all he does is type in his name U_i and password P_i; then the computer calculates $f(P_i)$ and compares it with the entry $(U_i, f(P_i))$. Agreement allows U_i access to the machine.

The intruder who wishes to break into the system and steal the time of U_i needs to find some X such that $f(X) = f(P_i)$. In other words, any intruder needs to be able to invert f, while our assumption of intractability makes this impossible.

A CONCRETE EXAMPLE

Suppose a password system is based on the weak one-way function described in Example 1 above. Each user generates two random k-bit integers m and n with k large, and these constitute his password. The security of the system is directly linked to the difficulty of factoring.

Example 2

A more secure system. Suppose that, instead of using k-bit integers, the system is based on each user generating two k-bit random primes p and q and using these as his password, with again

$$f(p, q) = pq.$$

This is much more secure. Factoring many of the $2k$-bit integers is very easy: half of them will be even and one third are divisible by 3. Factoring integers that are the product of two large primes 'is usually hard'.   □

This last example highlights an important property which a genuinely useful one-way function must possess. It must be 'hard to invert' on 'almost all' or a 'very large' proportion of the points in its range. Framing this into a mathematical definition is difficult, and we return to it in Section 13.3. For the moment, we work with our informal approach, reminding the reader that, as it stands, our definition of one-way is loose, and a function that we regard as one-way today could cease to be so tomorrow, e.g. if a fast factoring algorithm was found.

Exercises 10.1

1. Try factoring 518 940 557 on a microcomputer. Estimate how long the same machine would take to factor a 200-digit integer if it had multiprecision arithmetic facilities.
2. Assuming that taking products as in Example 2 is a one-way function, show that the following function f is also. The domain is the set of pairs of 2×2 matrices (A, B), each having integer entries that are n-bit primes, and $f(A, B) = AB$.

10.2   Using NP-hard problems as cryptosystems

At the moment, there is no fast (that is polynomial-time) algorithm known for any NP-hard problem, and if $NP \neq P$, no such algorithm exists. It is therefore a natural and attractive idea to build a cryptosystem around an NP-hard problem in such a way that cracking the cryptosystem would be equivalent to finding a fast algorithm for solving the particular problem.

This is the underlying idea behind one of the most famous—and certainly most controversial—cryptosystems of modern times, namely the Data Encryption Standard (DES). We will discuss this particular system in more detail in the next section. First, however, we describe the basic idea.

Consider the following computational problem.

ALGEBRAIC EQUATIONS OVER THE INTEGERS MODULO 2

Input Polynomials $p_1, \ldots, p_k$ in the variables $x_1, \ldots, x_n$, and with coefficients belonging to $\mathbb{Z}_2$

Question Do these polynomials have a common zero $(x_1, \ldots, x_n)$ in arithmetic modulo 2?

For example, the three equations

$$x_1 x_4 x_6 + x_2 x_4 x_5 - 1 = 0$$
$$x_1 x_2 + x_2 x_3 + x_3 x_4 - 1 = 0$$
$$x_1 x_3 + x_4 x_5 + x_1 x_6 - 1 = 0$$

have the solution $(1, 0, 1, 1, 1, 1)$.

Note that, since $x_i^2 = x_i$ in the set of integers modulo 2, then the polynomials feature no powers above the first, and hence we can represent these equations compactly in the form

$$(1, 4, 6), (2, 4, 5), 1; \quad (1, 2), (2, 3), (3, 4), 1;$$
$$(1, 3), (4, 5), (1, 6), 1:$$

Although the problem is easy in this particular case, it quickly becomes intractable when k and n are large, and we have the following result (see Garey and Johnson, 1979).

Theorem 1 *The problem of deciding whether algebraic equations modulo 2 have a solution is NP-hard.*

This particular NP-hard problem turns out to be the problem underlying the Lucifer cryptosystem designed by IBM and described by Feistel (1973). The system works as follows. First break the message M into blocks of size $2n$. Each block is treated by the same encryption procedure; so, with no loss of generality, we assume M is a block of length $2n$. Now divide M into two equal blocks so that $M = (M_0, M_1)$. The key K is a vector k which determines subkeys $k_1, \ldots, k_{d-1}$. For $2 \le i \le d$, we recursively define

(1)
$$M_i = M_{i-2} + f(k_{i-1}, M_{i-1}),$$

where f is some fixed nonlinear transformation and all algebra is in modulo-2 arithmetic.

The final cryptogram C is given by

(2)
$$C = e(M, K) = (M_{d-1}, M_d).$$

In order to recover M from C, all the decoder has to do is to invert (1) and iterate, so that

(3) $$M_{d-2} = M_d + f(k_{d-1}, M_{d-1}), \qquad M_{d-3} = M_{d-1} + f(k_{d-2}, M_{d-2}),$$

and so on, until we arrive at the original message $M = (M_0, M_1)$. Thus, provided that the decoder knows the form of f and the values of the keys, deciphering a cryptogram is as easy as encoding a message. Note also that, because of the reverse relationships (3), the function f need have no special properties such as being uniquely invertible.

In a typical situation, the adversary might know the form of f but not the keys. Whether or not he is able to recover the keys from a known plaintext–ciphertext pair depends on the form of f. If f is a straightforward linear transformation, then determining the key is an easy computational task. However, if f is a nonlinear transformation, say

$$f(x_1, \ldots, x_n; y_1, \ldots, y_n) = (p_1, \ldots, p_n),$$

where each p_i is a polynomial in $(x_1, \ldots, x_n, y_1, \ldots, y_n)$, then determining the key from a knowledge of M and C is easily seen to reduce to the problems of solving algebraic equations in the key variables. As we have seen, this is NP-hard. To illustrate this, consider the following very small example.

Example   Suppose that $n = 3$ and the encoding function f is given by

$$f(x, y) = f(x_1, x_2, x_3; y_1, y_2, y_3)$$
$$= (x_1 x_2 y_1 y_2, \; x_2 x_3 y_1 y_3, \; (x_1 + x_2) y_1 y_3).$$

Suppose also that there is an 8-bit key $k = (k_1, \ldots, k_8)$ and that the subkeys are given by

$$k_1 = (k_2, k_4, k_6), \qquad k_2 = (k_1, k_2, k_7),$$
$$k_3 = (k_3, k_5, k_8), \qquad k_4 = (k_6, k_7, k_8).$$

Then, if $M = (101\ 111)$, we get

$$M_0 = (101), \qquad M_1 = (111),$$
$$M_2 = M_0 + f(k_1, M_1) = (101) + (k_2 k_4, \; k_4 k_6, \; k_2 + k_4)$$
$$= (1 + k_2 k_4, \; k_4 k_6, \; 1 + k_2 + k_4),$$
$$M_3 = M_1 + f(k_1, k_2, k_7, M_2).$$

Thus, if we stop after only two iterations, so that

$$C = (C_1, C_2, C_3, C_4, C_5, C_6) = (M_2, M_3),$$

then

$$C_6 = 1 + (k_1 + k_2)(1 + k_2 k_4)(1 + k_2 + k_4),$$
$$C_5 = 1 + k_2 k_7(1 + k_2 k_4)(1 + k_2 + k_4),$$

and so on. □

Thus, in one sense, we have achieved the objective: namely a cryptosystem which has easy encryption/decryption processes but which forces the adversary trying to find the key by a chosen plaintext attack to face an NP-hard problem; hence (assuming that NP-hard really does mean intractable) the security of the system is assured.

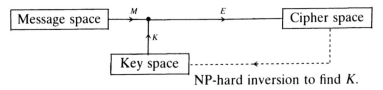

Fig. 1

However, there is a drawback here. Classical complexity theory as described in Chapter 9 is almost wholly concerned with worst-case behaviour. Thus, even though determining the key may be an NP-hard problem there is no guarantee that there are not a large number of 'easy inputs'.

Exercises 10.2

1. Suppose we take $n = 3$ and that the encoding function f is given by

$$f(x, y) = f(x_1, x_2, x_3; y_1, y_2, y_3)$$
$$= (x_1 x_2 y_1 y_2, \; x_2 x_3 y_3 y_1, \; (x_1 + x_2)y_1 y_3).$$

If $d = 3$ and the subkeys are

$$k_1 = (1,0,1) \qquad k_2 = (0,1,1),$$

find the cryptogram when the message $M = 111111$.

2. Suppose in the system described we take $n = 3$ so that our message M is a 6-tuplet. Suppose that the keys k_1, k_2, k_3, k_4 are just permutations of $(1, 2, 3)$ so that $f(k_i, M_i)$ just permutes the elements of M_i in accordance with the permutation k_i. If

$$k_1 = (1, 3, 2), \qquad k_2 = (2, 1, 3),$$
$$k_3 = (3, 2, 1) \qquad k_4 = (2, 3, 1),$$

show that the message 101011 is encrypted after four iterations as 011011. How is the receiver to decrypt 101011?

10.3   The Data Encryption Standard (DES)

The increase in computer communications and the advent of electronic fund-transfer systems in the early 1970s was one of the main factors behind the decision of the United States government to try to impose standards on the security and custody of Federal and other data banks. Accordingly, in 1973–4 the National Bureau of Standards (NBS) advertised for anyone interested to submit proposals for a data encryption standard. The hope/idea was that the encryption process might be fitted to a small chip, and that this would result in a mass-produced, widely used chip whose security was beyond question.

The scheme adopted was that proposed by IBM. It was based on their Lucifer scheme, and was of the same type as described in the last section. However, whereas the Lucifer scheme used a key of 128 bits, the key size in the scheme proposed to NBS was 64 bits, and eight of these bits were thrown away by the encryption algorithm. Based on this proposal the Data Encryption Standard (DES) was published in 1977 by the NBS for use by the Federal government to protect valuable and sensitive but unclassified data.

For the mathematical point of view, the DES can be regarded as a simple cipher with an alphabet of 2^{64} letters; that is, the message is broken up into blocks of 64 bits, and each such bit regarded as a letter.

The key K selected is one of the 2^{56} possible 56–bit words, and from it are abstracted deterministically 16 subkeys $K_1, \ldots, K_{16}$ where each subkey K_i is a 48-bit subword of the 56-bit K. Each of these subkeys K_i is used in a *standard building block* (SBB) to transform its 64 bits of input into 64 bits of output, these standard building blocks being used in series on the output as shown in Fig. 2. The initial permutation IP and its inverse IP^{-1} are of no cryptographic significance.

The way in which an SBB works is illustrated in Fig. 3.

The 64-bit input to the ith SBB is broken into a left half and right half (L, R). The block R becomes L' for the input to the $(i + 1)$th

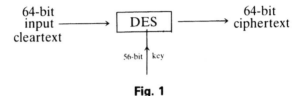

Fig. 1

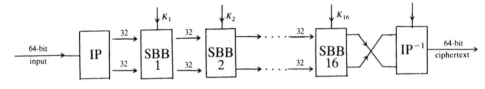

Fig. 2

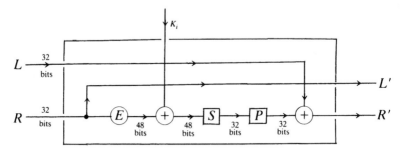

Fig. 3

SBB. The right half R' of the output is obtained by
(a) diffusing R using a diffuser/scrambler E to make it 48 bits,
(b) forming its modulo-2 sum with the key K_i,
(c) passing it through a nonlinear transformation S called an *S-box* to make 32 bits,
(d) permuting it by the operation P,
(e) summing it modulo-2 with L to get R'.

The crucial step of the enciphering is the nonlinear transformation S. Each S-box is a set of eight known transformations each of which transforms six bits of input into four bits of output.

Writing the overall transformation as

$$L' = R, \qquad R' = L + f(K, R),$$

we note that decryption can be carried out by using the keys in reverse order but with the same algorithm. Thus

$$R = L', \qquad L = R' - f(K, R) = R' - f(K, L').$$

The way in which the K_i are abstracted from the key K and the exact details of the transformations in the S boxes are public knowledge: they are specifically described in the Federal Register.

This fact—that everything is known about the encryption procedure except the individual key used—is at the same time the great strength and weakness of the system.

THE CONTROVERSY

It is fairly safe to assert that no other cryptographic device has aroused such a controversy as the Data Encryption Standard. M. E. Hellman and W. Diffie of Stanford University were the first to observe that the key length of 56 effective bits was uncomfortably short and suggested an increase to 64 bits, or even 128 bits as in the Lucifer scheme. As a result of this and an article by D. Kahn (New York Times, 3 April 1976), the NBS held two workshops to 'answer the criticisms'.

Discussion of the proposed standard was made difficult by the disclosure that some of the design principles used by IBM were classified and could not be divulged to members of the workshop.

The weight of the evidence was that the effective lifetime of the standard, if adopted, would be little more than 10 years. The standard was adopted in early 1977 and became effective in July of that year.

An interesting (but academic) question is whether the following statement of Davis (1978) in support of the standard is still true.

'Anyone buying cryptographic equipment which has been validated against the DES can be assured of a specific level of data security: namely that 2^{55} attempts and the use of the method of exhaustion are required to obtain any one key for the encryption algorithm used in the DES.'

In 1986 it was announced that the NSA is no longer prepared to approve the DES algorithm when it comes up for review as a Federal standard in 1988.

Exercise 10.3

1. If F^n denotes the set of functions from n-bit strings to n-bit strings, show that

$$|F^n| = 2^{n2^n}.$$

10.4 The discrete logarithm

In this section, we give another example of what is currently regarded as a one-way function. To do that, we need a small piece of elementary number theory. It is very easy to prove:

(1)   For any integer n, the set $\mathbb{Z}_n^*$ of integers modulo n that are coprime with n forms a group under multiplication modulo n.

The order of this group is denoted by $\phi(n)$ and is called the *totient* function or *Euler phi-function*. From (1), we know by Lagrange's theorem:

(2)   For any x $(1 \leq x < n)$ that is coprime to n,

$$x^{\phi(n)} = 1 \pmod{n}.$$

In the special case when n is a prime p, each integer less than p is coprime with p, so that

(3) $$\phi(p) = p - 1.$$

Thus we have a proof of the Euler-Fermat identity

(4) $$x^{p-1} = 1 \pmod{p} \quad (x = 1, \ldots, p-1).$$

Now consider any integer a in the range $1 \leq a \leq n$. If $\gcd(a, n) = 1$ and a has the additional property that $a^d \neq 1 \mod n$ for any $d, 1 \leq d < \phi(n)$, we say a is *primitive root* of n. For example, 2 is a primitive root of 5 because, in arithmetic modulo 5,

$$2^1 = 2 \quad 2^2 = 4, \quad 2^3 = 3, \quad 2^4 = 1.$$

Not every integer has a primitive root: 8 is the smallest such example, and the question of which integers do have primitive roots is contained in the following basic theorem from elementary number theory.

Theorem 1   (a) *The integer n has a primitive root if and only if n is 1, 2, or 4, or is of the form p^k or $2p^k$, where p is an odd prime.*
(b) *If n has a primitive root , it has precisely $\phi((\phi(n)))$ of them.*

Questions about how the primitive roots are distributed, which integers are primitive roots of some prime, and the like are fascinating unanswered questions in number theory. For example there is a famous long-standing (1927) conjecture of Artin:

Conjecture Every integer that is not a perfect square is the primitive root of some prime.

Of more direct interest to the applications we have in mind is the problem of finding primitive roots of an integer n. As far as we know, there is no algorithm known that, given an integer n of (say) N bits, will find a primitive root in time polynomial in N.

We can now define the discrete logarithm function. Take any

integer n that has a primitive root a. For any x $(0 \le x < \phi(n))$, if

(5) $y = a^x \pmod{n}$,

then x is called the *discrete logarithm* of y to the base a modulo n, and is written $x = \log_a y \pmod{n}$. The importance of a being a primitive element is that it guarantees that to any $y \in \mathbb{Z}_n^*$ there is a unique x, (see Exercise 2) so that the discrete logarithm is well defined.

We make the following claim:

(6)   The exponential function defined by (5) is a one-way function.

In other words, we are claiming exponentiation is 'easy' but that taking logarithms is 'hard'. One part of this assertion is easy to substantiate since it is easy to prove:

(7)   Exponentiation can be done in $2\lceil \log_2 n \rceil$ multiplications modulo n. In other words exponentiation belongs to P.

The idea behind the proof is repeated raising of powers, for example,

$$2^{18} = (((2^2)^2)^2)^2 2^2$$

can be obtained with just five multiplications.

At the moment, there is no known polynomial-time algorithm for computing discrete logarithms, and the difficulty of this computation is commonly regarded as on a par with that of factoring. We return to the current status of the difficulty of this problem in the last section of this chapter. For the moment, we consider it a 'hard' inversion problem.

Exercises
10.4

1. Show that 2^{75} can be obtained from 2 by just nine multiplications.
2. Prove that if a is a primitive root of n then the equation

$$y = a^x \pmod{n}$$

   has exactly one solution for x $0 \le x < \phi(n)$, given $y \in \mathbb{Z}_n^*$.
3. Show that, if a is not a primitive root of n, then the equation

$$y = a^x \pmod{n}$$

   may not have a unique solution for x given y.
4. Prove that, if a is a primitive root of n, then, for any integers $y, z \in \mathbb{Z}_n^*$

$$\log_a(yz) = \log_a y + \log_a z \pmod{\phi(n)}.$$

5. Prove that, if p and q are distinct primes and $n = pq$, then $\phi(n) = \phi(p)\phi(q)$.

10.5 Using the discrete logarithm to solve the key-distribution problem

One of the fundamental problems in classical cryptography is how to distribute the keys safely. If messages cannot be transmitted safely what guarantee is there that keys are secure?

In their seminal paper in 1976, Diffie and Hellman proposed a very elegant way of solving this problem. It hinges on the one-way nature of the discrete-logarithm problem and has now been implemented by Hewlett Packard and the Mitre Corporation. Consider a collection of users $(U_i : 1 \leq i \leq N)$ who wish to communicate with each other. Let p be a large prime (much larger than N) and let a be a primitive root of p.

A typical user U_i generates, independently of all other users, a pseudorandom number X_i in the range $\{1, \ldots, p-1\}$ and keeps this secret. However, he makes public, as his personal *public key*, the integer

(1)
$$Y_i = a^{X_i} \pmod p.$$

When users U_i and U_j wish to communicate privately, they use as their key

$$K_{ij} = a^{X_i X_j} \pmod p.$$

User U_i obtains K_{ij} by taking Y_j from the public file and then calculating K_{ij} by the formula

(2)
$$K_{ij} = Y_j^{X_i} \pmod p.$$

Similarly, U_j obtains K_{ij} by taking Y_i from the public file and using

(3)
$$K_{ij} = Y_i^{X_j} \pmod p.$$

Any other user, in order to obtain K_{ij}, must compute K_{ij} from the Y_i and Y_j which are public information.

If p is a prime slightly less than 2^b (say), then all quantities involved are representable as b-bit integers. Thus exponentiation takes at most $2b$ multiplications modulo p. In contrast, on our supposition that finding discrete logarithms is difficult, the enemy will need a superpolynomial number of operations to break into the system.

Example Three users A, B, C communicate privately using the above scheme. They choose as their common prime $p = 71$ and, as a primitive root, $a = 7$.

User A takes as his private key the integer 5, so that his public key is the integer

$$Y_A = 7^5 = 51 \pmod{71}.$$

User B has 12 as his private key and 4 as his public key so that, when A and B communicate, their key is the integer

$$K_{AB} = 4^5 = 51^{12} = 30 \pmod{71}.$$

User C, in order to find the communication key K_{AB}, has only the information $Y_A = 51$, $Y_B = 4$. Since the numbers here are deliberately chosen to be small, he has only to solve one of the equations

$$7^x = 51 \pmod{71} \text{ or } 7^x = 4 \pmod{71}. \qquad \square$$

The exact position of the security of the above key distribution system is not clear. It is clearly vulnerable to anyone able to compute discrete logarithms. However, as far as we know, it has not yet been proved that breaking the system is equivalent to computing discrete logarithms.

Exercises 10.5
1. In the key-distribution scheme described above, the common prime p is 11 and the primitive root used is $a = 2$. If user A has public key equal to 9, what is the private key of A?
2. In the scheme described in the previous example the user B has public key equal to 3. What is the key K_{AB} used by A and B when communicating with each other?

10.6 A cryptosystem with no keys

The key-distribution scheme described in the last section can be adapted to give a method of transmitting secret information over a public network. It is again based on the one-way nature of exponentiation and works as follows.

Suppose user A wishes to send a message to B. We suppose that the message can be represented as an integer M in the range $\{0, \dots, p-1\}$, where p is some large prime (this is easily done by breaking the message up into blocks). User A chooses an integer $a < p$ such that a is relatively prime to $p - 1$. User B chooses an integer $b < p$ which is also relatively prime to $p - 1$. The communication between A and B is then in three stages.

First, A sends to B the integer

(1) $$C = M^a \pmod{p}.$$

Secondly, B returns to A the integer

(2)
$$D = C^b \pmod{p}.$$

Finally, A computes the integer a' such that $aa' = 1 \pmod{p-1}$, and sends to B the integer

(3)
$$E = D^{a'} \pmod{p}.$$

The receiver B decrypts the message by the formula

(4)
$$F = E^{b'} \pmod{p},$$

where b' is such that $bb' = 1 \pmod{p-1}$.

There are various points to note. We must prove that the arithmetic 'works', in that $F = M$. We must also show that the required integers a' and b' exist and can be easily found. We deal with this first.

(5)
The integers a' and b' are well defined, since a and b were taken to be coprime with $p-1$, and hence are just the inverses of a and b in the group $\mathbb{Z}^*_{p-1}$.

(6)
Finding a' and b' given a and b is an easy computational task using Euclid's algorithm as described in Section 9.1. Since a is coprime with $p-1$, Euclid's algorithm will produce x and y such that $xa + y(p-1) = 1$. Take $a' = x$.

(7)
Now we prove that F is in fact the original message M. By definitions,
$$F = E^{b'} \pmod{p}$$
$$= (D^{a'})^{b'} = D^{a'b'} = C^{ba'b'} = C^{bb'a'} \pmod{p}.$$

But $bb' = 1 \pmod{p-1}$, so that
$$C^{bb'} = C^{1+t(p-1)} = C^{(p-1)t}C \pmod{p}.$$

Also, by the Euler–Fermat identity,
$$C^{p-1} = 1 \pmod{p}.$$

Thus we have
$$F = C^{a'} \pmod{p}.$$

But $C = M^a \pmod{p}$, so that
$$F = M^{aa'} \pmod{p}$$

and again, since $aa' = 1 \pmod{p-1}$, we use the Euler–Fermat identity to get the required result
$$F = M. \qquad \square$$

It can be shown that the above system is vulnerable to an attack from an enemy who can find discrete logarithms quickly.

This very clever scheme, which Konheim (1981) attributes to A. Shamir (unpublished), has of course the serious disadvantage of a 3-fold expansion in time.

Exercise   1. A small example of the above cryptosystem has $p = 31$. An enemy
10.6   intercepts

$$C = 4, \quad D = 4, \quad E = 2$$

being sent between users A and B. What was the message M being sent, and what were the private integers a and b of the two users?

10.7   On the difficulty of factoring and taking discrete logarithms

In this section, we aim to give some idea of the computational difficulty of two of the leading candidate one-way functions. Although the problem of finding algorithms for factoring large numbers was worked on by famous mathematicians such as Fermat and Legendre, it is in one of the problem areas in which a great deal of progress has been made over the last 15 years. The main reason for this is probably the advent of high-speed computing machines, though the importance of factoring algorithms in cryptographic circles is an 'economic' factor which should not be overlooked.

In contrast to the problem of testing primality, which (as we have seen) has an effectively polynomial-time algorithm, factoring appears to be much more difficult computationally. Several cryptosystems base their security on the presumed difficulty of the factoring problem and a great deal of activity and money seems to be being devoted to it.

First recall its formulation:

"Find a 'fast' algorithm that, given an integer N, will find integers x and y both greater than 1 such that $N = xy$."

The operative word here is 'fast'; in the language of complexity theory, our ideal would be an algorithm that, given any composite integer N of n bits would, in time polynomial in n, produce a factorization of N.

If we denote by $t(N)$ the running time $r(n)$ taken by the best general algorithm currently in use, then

$$r(n) = t(N) \simeq CN^{[(A \ln \ln N)/\ln N]^{\frac{1}{2}}},$$

where A is a constant >1. To see how far this is away from being a polynomial function of the number of bits $n = \log N$, note that

$$N^{[(A \ln \ln N)/\ln N]^{\frac{1}{2}}} = (2^n)^{[(A \ln n)/n]^{\frac{1}{2}}};$$

in other words, the running time is

$$r(n) \simeq c^{(n \log n)^{\frac{1}{2}}} \quad \text{for some constant } c > 1,$$

which is some way from being polynomial in n.

In practical terms, Riesel (1985) reports that, in 1985, factoring a 70-digit number on a supercomputer would (in the worst cases) take 10 hours, and estimates

$$t(10^{100}) \sim 1 \text{ year.}$$

Although this is out of reach at the moment, with the rapid improvement in technology (parallelism and the like), it does suggest that any cryptosystem based on the difficulty of factoring needs, in order to be really secure, to be using integers of the order of 400 bits.

This is particularly so, since not all large numbers are hard to factorize and certainly there have been notable successes in factoring numbers of special form. For example, the structure of the Mersenne numbers

$$M_n = 2^n - 1$$

has been completely determined for n up to 257 in terms of (at worst) under 10 hours on a CRAY I.

In a field which is as rapidly changing as this, any account is likely to be obsolete quickly. However, I hope it does give some indication of the state of the art (circa 1986). For those who would like to find out more, we recommend the surveys of Dixon (1984), Williams (1984) or the monograph of Riesel (1985).

HOW HARD IS DISCRETE LOGARITHM?

Finding 'fast' algorithms for obtaining discrete logarithms is a less well known computational problem than that of factoring. Nevertheless, because of its significance in cryptography, it too has received considerable attention in recent years.

Recall the problem: namely, given a, y, and N, find an integer x such that

$$a^x = y \pmod{N}.$$

An algorithm would be regarded as fast if it were polynomial in n, the number of bits of N, where usually N is a prime p or a prime power q.

When N is a prime p, there exists an algorithm developed independently by Western and Miller (1968), Pollard (1978), Merkle (1979), and Adleman (1979), which the last named shows has running time

$$r(n) = t(p) \simeq \exp[c(\ln N \ln \ln N)^{\frac{1}{2}}]$$
$$\simeq \exp[c(n \ln n)^{\frac{1}{2}}].$$

Hellman and Reynieri (1983) extended this to the case $N = p^m$ for p prime and extensions by Blake *et al.* (1984), Coppersmith (1984), and Odlyzko (1985) result in an algorithm with running time

$$r(n) \simeq \exp(cn^{\frac{1}{3}} \log^{\frac{3}{3}} n),$$

where $N = p^n$ and p is kept fixed while $n \to \infty$. The method proposed by Coppersmith just about makes possible computations in fields of size 2^{400}, but the method is not applicable to prime fields. Thus this is the reason for Coppersmith's assertion that computing discrete logarithms is easier over a field of size 2^m than over a prime field of about the same size.

To give some idea of the state of the art, Coppersmith in his 1984 paper reports the speed of these algorithms for taking logarithms in F_q with $q = 2^{127}$, which is a field in which the Diffie–Hellman key-exchange scheme had been implemented by Mitre Corporation. This is now insecure since Adleman's algorithm takes $\simeq 2$ weeks, the modification by Blake *et al.* about nine hours, and Coppersmith's algorithm 11 minutes all on an IBM 3081 K (a large main frame running at 16 million instructions per second). For a very good survey and detailed analysis of the problem see Odlyzko (1985).

It is interesting to compare the difficulty of discrete logarithm with that of factoring; several cryptosystems are based on one or other and the best existing algorithms for both problems seem to have running times which are asymptotically comparable.

At the time of writing, as far as I know, it is not known whether they are Turing equivalent, that is, that the existence of a polynomial-time algorithm for one implies the existence of a polynomial-time algorithm for the other.

PROBLEMS 10

1. If f and g are weak one-way functions with the same domain and range, which of the following are also weak one-way functions

    (a) $f + g$,   (b) fg,   (c) $f \circ g$?

2. If V_n denotes the set of binary vectors of length n and a map $f : V_n \to V_n$ is an *involution* if $f \circ f$ is the identity, prove that any involution must be 1–1.

3. Prove that the no-key cryptosystem is vulnerable to an attack from an enemy with a fast algorithm for the discrete logarithm.

4. Show that, if the keys of DES could be tested at the rate of one per $60 \, \mu s$, then testing all keys would take more than 68 000 years.

5. Let f be a one-way function. For any given constant c, show that f must 'hide' more than $c \log n$ bits, where n is the number of bits in the input to f, for all sufficiently large n.

6. A Feistel-type cipher as described in Section 2 operates on a block of $2n$ message symbols by the rule

$$M_{i+2} = M_i + AM_{i+1},$$

where A is an unknown nonsingular $n \times n$ matrix over the field of two elements.

Prove that this cryptosystem is vulnerable to chosen-plaintext attack.

7. Is it true that, if $f : S \to T$ can be computed in polynomial time, then $f^{-1} : T \to S$ can be computed in exponential time?

8. A Feistel-type cryptosystem as described above uses an iteration in which $f(k_i, M_i)$ is just the permutation of M_i determined by the key (permutation) k_i. Given the additional side information that the keys k_i are all equal to the same permutation π, prove that the system is vulnerable to a chosen-plaintext attack.

9. Show that there exists an algorithm for finding the discrete logarithm over the prime p that involves at most $2 \lceil p^{\frac{1}{2}} \rceil$ multiplications modulo p.
   (Pohlig and Hellman, 1978)

10. Explain how a cryptosystem that is secure against known-plaintext attack can be used to create a one-way function.

11. Suppose we encipher messages by the rule

$$C = M^K \pmod{p},$$

where p is a large prime, with $1 \le M \le p - 1$, and K is an integer with $1 < K < p - 1$.

Show that, if K is chosen to be coprime with $p - 1$, then the decryption algorithm

$$d(C) = C^D \pmod{p}$$

is correct in that, with $D = K^{-1} \pmod{p - 1}$, we have $d(C) = M$.
   (Pohlig and Hellman, 1978)

12. What happens to the cryptosystem above if we make an error and use an integer K that is not coprime with $p - 1$?

13. A modification of DES, which we call MDES, can be mathematically summarized as follows. It begins with f^{32}, which is a specification, for each key k of length 48, of a function $f_k^{32} \in F^{32}$. Given k and input x, it is easy to compute $f_k^{32}(x)$. Define g_k^{64} by

$$g_k^{64}(L, R) = \left(R, L + f_k^{32}(R) \right) \quad (|L| = |R| = 32).$$

Let h^{64} be a member of F^{64} defined by: h^{64} is 16 compositions of g_k^{64} with itself, using a different key in each composition, and all arithmetic is modulo 2. Thus h^{64} has a key length of $16 \times 48 = 768$. The transformation $h^{64}(x)$ is the MDES encryption of x. Prove that

(a) g_k^{64} is 1–1 onto and easy to invert if you know k.

(b) Given the 768-bit key, h_k^{64} is 1–1 onto and easy to invert.

[The difference between DES and MDES is that DES has a key of only 56 bits, which is used to generate (in a very simple way) the 16×48-bit key to be used in what is then effectively MDES.]

Note: Whether MDES is more secure than DES is not clear, though at the intuitive level it would appear much more so; for further details see Luby and Rackoff (1986).

14. Show that it is impossible for each of the distinct 768-bit keys in MDES to generate distinct encryption functions.

    Hint: Compare the number of keys with the number of distinct permutations, and use Stirling's approximation

$$n! \sim (2\pi n)^{\frac{1}{2}} n^n e^{-n}.$$

11

Public-key cryptosystems

11.1 The idea of a trapdoor function

The classical systems of cryptography all suffer from what is known as the *key-distribution problem*. This is the problem of establishing a private channel by means of which the sender and receiver of messages can exchange the key currently in use. Apart from the security risk, there is a severe practical problem in situations such as electronic mail systems, where communication needs to be rapid and cheap and where keys are frequently changed.

Diffie and Hellman addressed this fundamental problem in their seminal paper in 1976. In Chapter 10, we discussed one of their proposed solutions based on the intractability of the discrete-logarithm problem. Here we consider another of their suggestions, namely the idea of a public-key system that avoids the need for any secret communication of keys. This 'magic' system depends on the idea of what Diffie and Hellman call a 'trapdoor function' as we now explain.

All users in the system who wish to communicate with each other use the same encrypting algorithm e and the same decrypting algorithm d. Each user U_i has a pair of keys (K_i, L_i) such that, for any possible message M, we have the identity

$$d(e(M, K_i), L_i) = M,$$

where K_i is made public and stored on what is known as the *public file*; L_i is kept secret and is known as the *private key*; K_i is the *public key*. Hence, if another user U_j wishes to send U_i a message M, the procedure is as follows.

(a) User U_j must look up the public key K_i of user U_i in the public file.

(b) User U_j then sends the cryptogram

$$C = e(M, K_i)$$

to U_i across an open channel.

The security of the system depends on the use of functions e and d to encrypt/decrypt which possess the following properties.

Property 1: Given M and K it should be easy to compute $C = e(M, K)$.

Property 2: Given just the cryptogram C it is computationally not feasible to find M.

Property 3: Given the cryptogram C *and* the secret key L_i it is easy to determine the message M.

In other words, Properties 1 and 2 demand that the encrypting algorithm using the public key is a 'one-way function', but Property 3 demands that in addition there should be an 'inverse key', the possession of which makes it easy to invert. A one-way function with this additional property is called a *trapdoor function*.

For any such system to be practical, it needs the following further property.

Property 4: It should be easy to generate 'random' pairs of public/private keys (K_i, L_i).

In other words there should be 'enough' (K, L) pairs around for it to be impossible for any enemy to set up a 'look-up' table.

The adjective 'trapdoor' is now commonplace in cryptology and complexity theory as a generic description of a system or problem having the property that a computationally infeasible situation becomes easy to anyone in possession of certain 'trapdoor' information.

Having formulated this brilliant conception of a public-key cryptosystem, Diffie and Hellman seemed to have found some difficulty in coming up with detailed examples of such a system. While they hint at the possibility of basing systems on the one-way nature of exponentiation and the NP-completeness of the knapsack problem, it was left to Rivest, Shamir, and Adleman (1978) and Merkle and Hellman (1978) to produce the first explicit examples of public-key systems. These aroused a great deal of interest and we will discuss both in detail in the next two sections.

11.2 The Rivest–Shamir–Adleman (RSA) system

This is a public-key cryptosystem whose security is based on the belief that there is no fast way of factoring numbers that are the product of two large primes.

First we outline the encryption–decryption procedures. Details and proof will be described later.

THE ENCRYPTION–DECRYPTION PROCEDURE

I Find two 'large' primes p and q, and define n by

$$n = pq.$$

II Find a 'large' 'random' integer d that is relatively prime to the integer $(p - 1)(q - 1)$.

III Compute the unique integer e in the range $1 \le e \le (p - 1)(q - 1)$ from the formula

$$ed = 1 \pmod{(p - 1)(q - 1)}.$$

IV Make known the *public key*, which consists of the pair of integers (e, n).

V Represent M, the message to be transmitted, as an integer in the range $\{1, \ldots, n\}$; break M into blocks if it is too big.

VI Encrypt M into a cryptogram C by the rule

$$C = M^e \pmod{n}.$$

VII Decrypt by using the *private key* d and the formula

$$D = C^d \pmod{n}.$$

Example   The following small example is taken from Rivest, Shamir, and Adleman (1978).

Suppose we take as our pair of primes

$$p = 47, \qquad q = 59,$$

so that

$$n = pq = 2773.$$

We need d to be coprime to $(p - 1)(q - 1) = 2668$ and take

$$d = 157.$$

Using a variant of the Euclidean algorithm, $e = 17$. Then, with $n = 2773$ and encoding the alphabet in the form

$$\text{SPACE} = 00, \; A = 01, \quad B = 02, \; C = 03, \quad \ldots, \quad Z = 26,$$

the message

ITS ALL GREEK TO ME

is represented as

$$M \equiv 0920\ 1900\ 0112\ 1200\ 0718\ 0505\ 1100\ 2015\ 0013\ 0500.$$

Thus we will encipher the message as 10 blocks of 4 digits. The first block M_1 ($M_1 = 920$) will be enciphered by

$$C_1 = 920^{17} = 948 \pmod{2773}$$

and the whole message is enciphered as

$$0948\ 2342\ 1084\ 1444\ 2663\ 2390\ 0778\ 0774\ 0219\ 1655.$$

The reader may check that deciphering works, for example $948^{157} = 920 \pmod{2773}$ etc.   □

THE UNDERLYING MATHEMATICS

We have to show that the encryption–decryption procedure actually works, in the following sense; with a given public key (e, n) and private key d, and for any message M represented by an integer in the range 0 to $n-1$, we have that, if

(1) $$e(M) = C = M^e \pmod{n},$$

then

(2) $$d(C) = C^d = M \pmod{n}.$$

Proof of (2): Consider the decrypted message

$$D = C^d \pmod{n}.$$

Then, since $C = M^e \pmod{n}$, we have

$$C = M^e - sn,$$

where s is some non-negative integer. Hence

$$D = (M^e - sn)^d \pmod{n}$$

and, expanding this by the binomial theorem, we get

$$D = M^{ed} \pmod{n}.$$

But, by definition, $ed = 1 \pmod{(p-1)(q-1)}$. Hence

$$D = M^{t(p-1)(q-1)+1} \pmod{n},$$

where t is some non-negative integer.

We now use the following easy lemma.

Lemma   *For any pair of distinct primes p and q, and any pair of positive* *integers (x, u), if $x^u = x$ $(\bmod\, p)$ and $x^u = x$ $(\bmod\, q)$, then*

$$x^u = x \ (\bmod\, pq).$$

Proof $x^u - x$ is divisible by p and q and therefore by pq. $\square$

Now, since p is prime and provided that $M \neq 0$ $(\bmod\, p)$, the Euler–Fermat identity gives

$$M^{p-1} = 1 \ (\bmod\, p).$$

Hence, for some integer r,

$$M^{p-1} = rp + 1$$

and thus

$$M^{t(p-1)(q-1)+1} = (rp + 1)^{t(q-1)}M$$
$$= M \ (\bmod\, p).$$

Since the statement holds trivially in the case $M = 0$ $(\bmod\, p)$, we know it for all M. By identical arguments,

$$M^{t(p-1)(q-1)+1} = M \ (\bmod\, q)$$

and thus, by the above lemma,

$$M^{t(p-1)(q-1)+1} = M \ (\bmod\, pq),$$

as required. $\square$

THE 'OBVIOUS' METHODS OF CRACKING THE RSA SCHEME

Recall now the position of someone who wishes to break into the system. We must assume that the following are known:

(3) n and e (obviously since this is the public key),

(4) unlimited pairs of messages M and corresponding cryptograms C, each (M, C) pair being two integers in the range 1 to $n - 1$.

Possible ways of cracking the system are:
(a) *Factoring n*:
Obviously, if an enemy could factor n, he could determine $\phi(n) = (p - 1)(q - 1)$ and hence the private key d.
(b) *Computing $\phi(n)$* without factoring n.
If $\phi(n)$ is known, then clearly it is easy to find d. However, we can easily see that knowledge of $\phi(n)$ leads to an easy way of factoring n.

This is because of the identities

$$p + q = n - \phi(n) + 1, \qquad (p - q)^2 = (p + q)^2 - 4n,$$
$$q = \tfrac{1}{2}[(p + q) - (p - q)].$$

Alternatively, a sceptic might argue that the decipherment problem is just the problem of computing eth roots modulo n, and is not a well-known difficult problem like factoring. Rivest, Shamir, and Adleman (1978) conjecture that any general method of breaking their scheme yields an efficient algorithm for factoring. This would establish that any method of cracking the RSA scheme was as difficult as factoring. As yet, this conjecture is unproven.

IMPLEMENTING THE PROCEDURE

In addition to verifying that the procedure above works in the sense that the deciphered message is the original message sent, we also have to ensure that the method is practical. This means that we have to show that, with choices of p, q, and d so large that security is assured, the processes of enciphering and deciphering can be done quickly enough to make the scheme viable.

If we analyse more closely the steps I–VII of the encryption–decryption process we note

(5) Finding large primes p and q can be done quickly by a trial-and-error procedure using the Rabin–Solovay–Strassen algorithm described in Section 9.6.

(6) Finding an integer d coprime with $(p - 1)(q - 1)$ can be done by taking d prime and larger than $\max\{p, q\}$.

(7) To find e such that

$$ed = 1 \pmod{(p - 1)(q - 1)},$$

we use Euclid's algorithm shown in Section 9.1. Since d is coprime with $\phi(n) = (p - 1)(q - 1)$, we get, in polynomial time, integers a and b such that

$$ad + b\phi(n) = 1,$$

and take $e = a$.

(8) The actual encryption and decryption steps VI and VII are then obtained by exponentiation modulo n and, as shown in Section 10.4, this can be done in polynomial time.

SECURE CHOICES OF p, q

On the assumption that the RSA scheme is as hard as factoring and with the current state of the art in factoring, a choice of p and q as primes of the order of 100 decimal digits would appear reasonably safe.

A further precaution is to ensure that $p - 1$ and $q - 1$ do not have many small prime factors. This reduces the risk of n being factored; see also Problem 13.

Exercises 1. In a public-key cryptosystem using the RSA method, you intercept the
11.2   cryptogram $C = 10$ sent to a receiver whose public key is $e = 5$ and $n = 35$.
                 Find the integer M which was sent.
          2. In an RSA system the public key of a given user is $e = 31$ and $n = 3599$.
                 What is the private key of the user?

11.3 Knapsack-based systems

The other very early (1978) public-key system was that introduced by Merkle and Hellman based on what is commonly described as the knapsack problem. More accurately, the underlying computational problem is the conceptually very simple problem known as SUBSET SUM. This is defined as follows:

SUBSET SUM
Input: Positive integers $a_1, a_2, \ldots, a_n, t$
Question: Is there a subset $J \subseteq \{1, \ldots, n\}$ such that

$$\sum_{i \in J} a_i = t?$$ □

This problem is one of the classic NP-complete problems.

As an illustration of its computational difficulty consider how long it would take you to verify the following:

Example (Hellman, 1978). The integer 516 has no representation as the sum of integers from the set

$$\{14, 28, 56, 82, 90, 132, 197, 284, 341, 455\}.$$

The integer 515 on the other hand has three distinct representations.   □

The Merkle–Hellman public-key system, which is based on the difficulty of this problem, works as follows.

ENCRYPTION

(1) The message to be sent is encoded in binary form m.
(2) The public keys are a collection of n-tuples $(a_1, \ldots, a_n)$ of positive integers.
(3) The binary message m to be sent is broken up into n-blocks so that $m = m_1 \ldots m_t$ where each m_j is an n-tuple of zeros and ones.
(4) For each j $(1 \leq j \leq t)$, set

$$c_j = \sum_{i=1}^{n} M_i a_i$$

where $m_j = (M_1, \ldots, M_n)$
(5) Transmit the sequence $c_1, \ldots, c_t$ of positive integers as the enciphered message.

DECRYPTION

On the surface, the receiver and any interceptor of the cryptogram $c_1 c_2 \ldots c_t$ are faced with the same problem; in order to decipher the message from the $c_1, \ldots, c_t$ and the public key $(a_1, \ldots, a_n)$ they have to solve t different NP-hard problems, one for each c_i.

The Merkle–Hellman system is based on the fact that not all instances of the NP-complete problem are hard to solve. We start with a definition.

Call a sequence $a_1, \ldots, a_n$ *superincreasing* if for each k $(1 \leq k \leq n-1)$

(1)
$$a_{k+1} > \sum_{i=1}^{k} a_i.$$

It is very easy to prove the following result.

Lemma   *There is a fast (polynomial time) algorithm for solving the class of SUBSET SUM problems in which the input $(a_1, \ldots, a_n)$ is superincreasing.*

Proof   Suppose $a_1, \ldots, a_n$ is superincreasing and we need to represent t as
(Sketch)   the sum of a subset of the a_i.

If r is the greatest i such that $a_i \leq t$, then we know that

$$t = a_r + s,$$

where we now need to find a representation of s as the sum of the superincreasing sequence $(a_1, \ldots, a_{r-1})$. Repeating this procedure gives our representation of t. $\square$

The basis of the Merkle–Hellman system is therefore:
(I) A typical user A selects an 'easy' superincreasing sequence of integers $(e_1, \ldots, e_n)$.
(II) User A then chooses a pair of 'big' coprime integers w and N and transforms the chosen easy sequence $(e_1, \ldots, e_n)$ into a 'hard' sequence $(T(e_1), \ldots, T(e_n))$ by the rule

$$T(e_i) = we_i \pmod{N} \quad (1 \le i \le n).$$

This transformed vector $(T(e_1), \ldots, T(e_n))$ becomes A's *public key*.

Note: N has to be bigger than $e_1 + e_2 + \ldots + e_n$.

Suppose now that this user A receives a cryptogram c sent using his public key $(T(e_1), \ldots, T(e_n))$. We claim decipherment is easy because of the following lemma.

Lemma *If c is the cryptogram sent to user A using A's 'hard' public key $(T(e_1), \ldots, T(e_n))$, then the cryptogram c' that would have been sent using the easy key $(e_1, \ldots, e_n)$ is given by*

$$c' = w^{-1}c \pmod{N}.$$

Proof Write $a_i = T(e_i)$ so that, if $M = (M_1 M_2 \ldots M_n)$ is the message, then

$$c = M_1 a_1 + \ldots + M_n a_n.$$

But

$$a_i = we_i + d_i N \quad (1 \le i \le n),$$

where d_i is an integer. Hence

$$c = \sum M_i a_i = \sum M_i we_i \pmod{N}$$

so that

$$w^{-1}c = \sum M_i e_i \pmod{N}$$

as required. $\square$

We illustrate the method by a very small example.

Example   Suppose we take as our hidden easy sequence

$$(1, 3, 5, 11, 35),$$

and then use as our private key

$$w = 5, \qquad N = 53.$$

Thus the public key is

$$(5, 15, 25, 2, 16).$$

If another user B of the system wishes to send me a message, say

$$M = 10011010101101101110,$$

the first step of B is to break M up into blocks of 5:

$$M = 10011 \ 01010 \ 11011 \ 01110.$$

The encoding of M is now as follows.

$$10011 \mapsto 5 + 2 + 16 = 23$$
$$01010 \mapsto 15 + 2 = 17$$
$$11011 \mapsto 5 + 15 + 2 + 16 = 38$$
$$01110 \mapsto 15 + 25 + 2 = 42$$

so that the cryptogram sent will be

$$C = (23, 17, 38, 42).$$

Decoding is easy: just multiply each integer by the inverse $w^{-1} = 32$ to get the decoding of 23 as

$$23 \mapsto 23.32 = 47 \pmod{53}$$
$$= 35 + 11 + 1$$
$$= 10011 \text{ as required.} \qquad \square$$

Because this public-key system is based on a provably hard problem, it might have been thought to be less susceptible to cracking than (say) the RSA. After all, factoring has not been shown to be NP-hard. However, this is not true. In 1982, Shamir produced an analysis which showed the vulnerability of the Merkle–Hellman system. He proved that 'almost all' instances could be solved in polynomial time. This was followed by an attack by Adleman (1983) which gave strong evidence that iterated versions of the Merkle–Hellman scheme can be broken.

As far as practical implementations are concerned, the confidence of the user is essential; hence, in the light of the Adleman–Shamir results, no system based on the Knapsack problem would seem to have a viable commercial or practical future. Nevertheless, for historical reasons, and for its role in giving an understanding of what exactly is needed in a public key system, we have included it in this course in a more prominent position than its practical significance now warrants.

Exercises 1. A user A of the Merkle–Hellman knapsack scheme has a public key
11.3   consisting of the sequence of integers

$$(37, 25, 13, 20, 3).$$

His private key is the pair $w = 37$, $N = 43$. What is the easy sequence?

2. If the letters of the alphabet are encoded in binary so that

$$A \mapsto 00000, \quad B \mapsto 00001, \quad \ldots, \quad Z \mapsto 11001,$$

what word is being sent in the previous exercise if the cryptogram is the sequence

$$c = (20, 50, 40, 13)?$$

11.4 A public-key system as intractable as factoring

As we mentioned in our discussion of the RSA system, we know that it could be cracked if an efficient algorithm for factoring existed. In the language of complexity theory

$$\text{CRACKING RSA} \propto \text{FACTORING}.$$

However, it has not been *proved* that it is actually as hard as factoring.

A different number-theoretic idea due to Rabin (1979) gives a public key system which is provably equivalent to factoring. We now describe this system. Each user of the system chooses a pair (p, q) of large distinct primes which he keeps secret. He also chooses an integer $B < N = pq$.

The public key is the pair (B, N).

The *private key* is the factoring (p, q) of N.

The *encryption function e* of a message M, with M (blocked if necessary) represented as an integer in the range $\{1, \ldots, N-1\}$, is

(1) $$e(M) = M(M + B) \pmod{N}.$$

If C is the resulting cryptogram, the decoding problem is to find M such that

(2) $$M^2 + BM = C \pmod{N}.$$

The key idea to the decryption algorithm is the following proposition.

Lemma 1   *A solution to the congruence*

(3) $$x^2 + Bx = C \pmod{pq}$$

can be obtained by taking solutions u and v to the congruences

(4) $$u^2 + Bu = C \pmod{p},$$

(5) $$v^2 + Bv = C \pmod{q},$$

and integers a and b satisfying

(6) $$a = 1 \pmod{p}, \quad a = 0 \pmod{q}, \quad b = 0 \pmod{p}, \quad b = 1 \pmod{q},$$

and then

$$x = au + bv$$

satisfies (3).

Proof   Routine checking and writing $a = 1 + kp$ or $a = lq$ as needed.   □

We now note the following result.

Lemma 2   *Since p and q are prime, integers a and b satisfying* (6) *can be found using a variant of Euclid's gcd algorithm in time polynomial in $\log pq$.*

Proof   Just carry out Euclid's algorithm to find the gcd of p and q; since they are coprime, this will give a representation of $1 = ep - fq$. Now take $a = fq$ and $b = ep$.   □

Thus our decryption algorithm will be complete if we can solve the congruence modulo prime integers. But this is straightforward.

Lemma 3   *Solving*

$$u^2 + Bu = C \pmod{p}$$

is equivalent to solving

(7) $$y^2 = C + (4^{-1})_p B^2 \pmod{p},$$

where $(4^{-1})_p$ denotes the multiplicative inverse of 4 in the integers modulo p.

Proof $(4^{-1})_p$ exists since p is prime and the conclusion (7) follows by completing the square.   □

Since q is also prime, the statement of Lemma 3 holds with q replacing p, and hence we have reduced the decryption to the problem of finding square roots modulo primes.
So let us consider the problem of solving

(8) $$y^2 = d \pmod p.$$

First note that, in general, this may not have a solution. The integers d for which (8) has a solution are called *quadratic residues* modulo p, and a fundamental result is the following, known as Euler's Criterion.

Theorem   *An integer a in the range $1 \le a \le p - 1$ is a quadratic residue modulo*
1   *an odd prime p if and only if*

(9) $$a^{\frac{1}{2}(p-1)} = 1 \pmod p.$$

The proof can be found in any basic text on number theory.
We can use this result to prove the next one.

Lemma   *If p is a prime of the form $4k - 1$, and d is a quadratic residue modulo*
4   *p, a solution of the congruence*

$$y^2 = d \pmod p$$

is given by

$$y = d^k \pmod p.$$

Proof   Since we know d is a quadratic residue modulo p, Euler's criterion gives

$$d^{\frac{1}{2}(p-1)} = 1 \pmod p.$$

Since $k = \frac{1}{4}(p + 1)$, we have

$$d^{\frac{1}{4}(p+1)} d^{\frac{1}{4}(p+1)} = d^{\frac{1}{2}(p+1)} = d^{\frac{1}{2}(p-1)} d$$
$$= d \pmod p.   □$$

Thus, by combining these lemmas, we have proved the following assertion.

(10)   Provided that the primes p and q are both congruent to 3 modulo 4, then the decryption procedure can be carried out in polynomial time.

Proof   The receiver, who knows the factors p and q of n, also knows the received cryptogram must be a quadratic residue, and can solve the congruences modulo p and q and then use Lemmas 1 and 3 to obtain the solution M to (2). ☐

Rabin has actually proved more than (10), by showing that, even when p and q are not of this special form, the congruences modulo p and modulo q can be solved by a random-polynomial-time algorithm. This is harder to prove, and we refer to his paper for details. However, since we regard a randomized polynomial-time algorithm as effective, we can sum up as follows.

(11)   For any choice of primes p and q, the decryption procedure can be carried out by a randomized algorithm working in polynomial time.

We illustrate with a small example.

Example   Suppose A takes as his public key

$$B = 2, \qquad N = 77,$$

so that his private key is the factorization

$$p = 7, \qquad q = 11.$$

If the message M is 3, then

$$C = M^2 + 2M = 15 \pmod{77}$$

Hence, to decrypt, A has to solve

$$u^2 + 2u = 15 = 1 \pmod 7, \qquad v^2 + 2v = 15 = 4 \pmod{11}.$$

The solutions are given by solving

$$(u + 1)^2 = 2 \pmod 7, \qquad (v + 1)^2 = 5 \pmod{11},$$

to get

$$u + 1 = \pm 2^2 = \pm 4 \pmod 7, \qquad v + 1 = \pm 5^3 = \pm 4 \pmod{11}.$$

Hence

$$u = 3 \quad \text{or} \quad 2, \qquad v = 3 \quad \text{or} \quad 6.$$

Then, since Euclid's algorithm gives $a = 22$ and $b = 56$ as integers satisfying condition (6) with $p = 7$ and $q = 11$, the solution to

$$x^2 + 2x = 15 \pmod{77}$$

is given by

$$x = \begin{cases} 3 \cdot 22 + 3 \cdot 56 \\ 3 \cdot 22 + 6 \cdot 56 \\ 2 \cdot 22 + 3 \cdot 56 \\ 2 \cdot 22 + 6 \cdot 56 \end{cases} \pmod{77}.$$

That is, the receiver A has four possible interpretations of the cryptogram C, namely

$$M_1 = 3, \qquad M_2 = 17, \qquad M_3 = 58, \qquad M_4 = 72. \qquad \square$$

This example highlights a practical disadvantage of the Rabin scheme, namely the receiver usually has four possible interpretations from which to choose the message. Usually this can be avoided by knowledge of side information; for example, with large N, the binary interpretation must decode back into English words.

Any adversary, on the other hand, who does not know the factors p and q of N is faced with a much more difficult problem. That this is really so follows from the second theorem of Rabin, which we now state.

Theorem 2 *If, for a fraction $1/\log N$ of the integers d for which there is a solution to*

$$y^2 = d \pmod{N},$$

we could find one such y, then we could factor N in random polynomial time.

Since we have seen in Section 9 that any problem known to be solvable in random polynomial time can be efficiently solved, we can restate the above theorem informally as follows.

Theorem 2′ *Cracking the Rabin public-key system is equivalent to finding an efficient algorithm for factoring.*

Sketch Proof The idea behind the proof is the following observation.

(12) If $x, y \in \mathbb{Z}_N$, the integers modulo N, are such that

$$x^2 = y^2 \pmod{N}, \qquad x \neq \pm y \pmod{N},$$

then the gcd of $x + y$ and N is a prime factor of N.

Proof $x^2 = y^2 \pmod{N}$ implies $x^2 = y^2 + rN$, where $r \in \mathbb{Z}$. Hence

$$(x - y)(x + y) = rN,$$

so that the gcd of $x + y$ and N is a prime factor of N. $\square$

Hence, suppose we have an algorithm $\mathcal{A}$ which, when given (q, N), for a fraction $1/\log N$ of the q's, gives as output one square root of q

modulo N. Then we can factor N by iterating the following step: Select z at random in $\mathbb{Z}_N$, with gcd $(z, N) = 1$, and compute $q = z^2 \pmod{N}$. Feed the algorithm $\mathcal{A}$ with the input (q, N). If $\mathcal{A}$ gives as output a square root of q different from z or $-z$ modulo N, then (12) shows that we can factor N. The expected number of iterations is low since, by the hypothesis, there is a $1/(2 \log N)$ chance of factoring N at each iteration. $\square$

**Exercises
11.4**

1. User A of a Rabin cryptosystem has public key

$$B = 28, \qquad N = 143.$$

   An enemy intercepts the cryptogram $C = 73$ which is being sent to A. What are the possible messages being sent?

2. Two users A and B of a Rabin cryptosystem have public keys (a, N) and (b, N) respectively.
   A third user U sends a message M to A by first sending it to B, who then transmits it to A. If $b = 0$, show that an enemy can recover M from intercepting the cryptogram C from B to A by solving the congruence

$$M^4 + aM^2 = C \pmod{N}.$$

11.5   A public-key system based on the discrete logarithm

Roughly speaking, factoring integers and computing discrete logarithms seem to be problems of comparable difficulty. The public-key systems based on the difficulty of factoring are well known. In 1985, T. Elgamal announced a system based on the apparent intractability of the discrete logarithm.

It is closely related to the key-distribution algorithm described in Section 10.5, and works as follows.

First, all users of the system are informed of a 'large' prime p together with a primitive root a modulo p. Thus a is an integer less than p, and

$$a^{p-1} = 1 \pmod{p},$$

but

$$a^{d-1} \neq 1 \pmod{p}$$

for any d, with $1 < d < p$.

The *private key* of a user B of the scheme is an integer x_B chosen at random by B with

$$1 \leq x_B \leq p - 1.$$

The *public key* of B is the integer

$$y_B = a^{x_B} \pmod{p}.$$

Suppose now that A wishes to send B a message M, where, as usual, we can regard M as an integer satisfying $1 \leq M \leq p - 1$.

In order to encrypt M, sender A proceeds as follows:

(i) A chooses at random some integer k such that

$$1 \leq k \leq p - 1.$$

(ii) A computes the 'key'

$$K = y_B^k \pmod{p}.$$

(iii) A encrypts the message M as the pair of integers (C_1, C_2), where

$$C_1 = a^k \pmod{p}, \qquad C_2 = KM \pmod{p}.$$

Thus the encryption process e is a map

$$e : M \mapsto C = (C_1, C_2)$$

which doubles the length of the message.

Note   Unlike previous schemes discussed, this approach has the property that, because k was chosen at random, if A was sending M a second time, then it would be most unlikely to have the same encoding.

In order to decode a received cryptogram (C_1, C_2), the receiver B proceeds as follows.

Decoding Algorithm
Step 1: Recover K by the rule

$$K = y_B^k = a^{x_B k} = (a^k)^{x_B}$$
$$= C_1^{x_B} \pmod{p},$$

where all arithmetic is modulo p. This is easy for B since x_B is his private key.
Step 2: Recover M by dividing C_2 by K (again in arithmetic modulo p). ☐

Example   As a very small example, suppose that the chosen prime p is 71 and its primitive root a is 7.

If B has public key $y_B = 3$, and A chose the 'random' key $k = 2$ then

$$K = 9,$$

and hence the encrypting of $M = 30$ would be

$$30 \mapsto (49, 57).$$ □

From the above description, it is easy to see that the scheme described is vulnerable to any attack having a fast algorithm for computing discrete logarithms.

As yet, it is not known whether breaking the system is equivalent to computing discrete logarithms. However, Elgamal does show that breaking the scheme is equivalent to breaking the Diffie–Hellman key exchange scheme described in Section 10.5.

Exercises 1. Prove the assertion that a fast algorithm for computing discrete logarithms
11.5   would break the above scheme.
     2. In the Example above, if A chooses a different key so that the encoding of 30 is

$$30 \mapsto (59, C_2),$$

what is the integer C_2?

11.6 Error-correcting codes as public-key systems

Edgar Allan Poe, who rather fancied himself as a cryptanalyst, once issued the challenge in a regular column to a Philadelphia periodical *Alexander's Weekly Messenger* that he could solve 'forthwith' any cryptogram ('mono alphabetic substitution cipher'), He was taken up on this challenge by one G. W. Kulp who submitted a 43-word ciphertext. Poe showed in a subsequent column that the challenge cipher text was 'a jargon of random characters having no meaning whatsoever'. In 1975, Kulp's cryptogram was cracked by B. J. Winskel and M. Lyster (see Gardner, 1977). In addition to a major error by Kulp, there are no less than 15 minor (? printing) errors, and the reason for its difficulty becomes apparent.

Whether intentional or not, the introduction of errors certainly confuses the enemy cryptanalyst, and this forms the basis of an ingenious public-key scheme proposed by R. J. McEliece (1978). In broad outline, the scheme works as follows.

A typical user, Alice, has as her *public key* a $k \times n$ binary matrix $\hat{G}$; if Bob wishes to send a binary message M to Alice, he uses the following encoding algorithm.

(I) Break the message into blocks of size k, and deal with each block separately.

(II) Encode each block m by

(1)
$$c = e(m) = m\hat{G} + z,$$

where z is a random *error* vector of length n and weight $\leq t$, and t is some previously agreed integer.

The *private key* of Alice is a pair of matrices (S, P), where S is a $k \times k$ *scrambler* matrix (nonsingular) and P is a random *permutation matrix*. The role of S and P is that

(2)
$$\hat{G} = SGP,$$

where G is the $k \times n$ generator matrix of a linear code which can correct up to t errors. It is now clear how the system works. The decryption process of Alice is:

(III) Form

$$\hat{c} = cP^{-1}$$
$$= (mSGP + z)P^{-1}$$
$$= mSG + zP^{-1}.$$

(IV) Since z has been chosen to have weight $\leq t$ and the code generated by G will correct up to t errors, the decoding algorithm for the linear code generated by G will decode $\hat{c}$ into mS.

(V) Unscramble mS by multiplying by S^{-1}.

McEliece's proposal was to use a *Goppa code* as the underlying basis. These are a family of good linear codes with a very efficient decoding algorithm (Patterson, 1975). Their parameters are of the form

$$n = 2^a, \quad k = n - at, \quad d \geq 2t + 1,$$

for a and t given integers.

Thus, with $n = 1024 = 2^{10}$ and $t = 50$, so that $k = 524$, the enemy is faced with resurrecting a message from text with up to 50 random errors in each word of length 1024.

PROBLEMS 11

1. When using the RSA scheme, Alice sends message M to Bob whose public key is $e = 43$, $n = 77$. If the cryptogram intercepted is $C = 5$, what is M?

2. A user of the public-key system based on the knapsack problem inadvertently chooses as his private key the integers 3 and 2^{N+1} and

takes as his 'easy' set the collection of integers $\{1, 2, 2^2, 2^3, \dots, 2^N\}$. Show that any message sent to this user is easy to decode, even without knowledge of the private key.

3. Show that, in the RSA and Rabin public-key schemes, the encryption of N bits of cleartext into N bits of ciphertext takes $O(N^3)$ bit operations.

4. Prove that any algorithm which solves the general knapsack problem with $(a_i : 1 \le i \le n)$ arbitrary integers must have time complexity at least $O(n)$. What is the order of magnitude of the time complexity of an exhaustive search algorithm for the problem?

5. If the superincreasing sequence used in the knapsack scheme contains only n integers, and the private key is (w, N), show that, when $n > \log_2 N$, there exist distinct integers x and y which have the same encoding.

6. Show that any sequence of integers $(a_i : 1 \le i \le n)$, with $a_i = q^{b_i}$, is superincreasing when the sequence $(b_i : 1 \le i \le n)$ is monotone increasing.

7. Suppose that a user of the RSA scheme chooses by mistake a large prime for his modulus N. Show that in this case decryption is easy.

8. The Chinese remainder theorem says that, if n_1 and n_2 are coprime, then there is exactly one solution x $(0 \le x < n_1 n_2)$ such that

$$x = a_1 \ (\text{mod } n_1), \qquad x = a_2 \ (\text{mod } n_2).$$

(i) Prove this.
(ii) In Rabin's public key scheme, Alice sends the same message M to Bob (B) and Charles (C) so that the encryption is

$$M^2 \ (\text{mod } n_\text{B}) \quad \text{and} \quad M^2 \ (\text{mod } n_\text{C}).$$

Using the Chinese remainder theorem, show that, when n_B and n_C are coprime, the enemy can recover M.

9. Let p and q be distinct primes, with $n = pq$.
(i) Given that d and n are coprime, show that if

$$x^2 = d \ (\text{mod } n)$$

has one solution then it has four solutions.
(ii) Give an example to show that this need not be true if d and n are not coprime.

10. Show that in Elgamal's public-key scheme based on the discrete logarithm, the enciphering procedure requires about $2 \log p$ multiplications modulo p.

11. Suppose that, to save space and time, all users of a public-key system based on RSA use the same modulus N, which is chosen by a central authority who also distributes the (e_i, d_i) pairs to each user. Show that knowledge of one (e, d) pair with

$$ed = 1 \ (\text{mod } \phi(N))$$

enables one to factorize N by a randomized algorithm, and hence crack the entire system. (This is called the common-modulus protocol).

(De Laurentis, 1984)

12. Consider the RSA scheme with modulus N. An integer M, $1 \le M \le N-1$ is a *fixed point* if it is encrypted into itself. Prove that if M is a fixed point then so is $N - M$.

13. Show that the RSA scheme with parameters p, q, e, d has $r + s + rs$ fixed points M, $1 \le M \le N - 1$, where

$$r = \gcd(p - 1, e - 1),$$
$$s = \gcd(q - 1, e - 1).$$

[Since encryption schemes with many fixed points are inherently unsafe, it is desirable to choose p, q to make r, s low.]

12

Authentication and digital signatures

12.1 Introduction

The problem of authenticating the source of a message is the underlying problem of this section. It is a problem of increasing importance in its applications; for example, as electronic mailing systems rapidly replace existing paper mailing systems for business transactions, it can be essential that the recipient of a message has proof that the message originated with the supposed sender.

There are various versions of this problem. Suppose Alice (A) is communicating with Bob (B). In an *authentication* problem, the receiver Bob has only to convince himself that the message received came from Alice. In a *signature* problem, the receiver Bob must be able to 'convince a judge' that the message originated with Alice. In other words, he needs a 'proof' by which he can demonstrate to an impartial referee that Alice did send the message.

The importance of authentication is highlighted in a recent *Newsweek* article on electronic office mail. In one incident on April Fools' Day, an employee sent around a bogus electronic memo to his co-workers saying that their plant was to close for two weeks. Rumours of the impending shutdown were taken so seriously that everyone stopped work to talk about it, resulting in a $1 million productivity loss. The same article reports the perennial problem of privacy. Although electronic mail is supposed to be secret, there is the much-quoted tale of a frank memo about an employee's incompetence being misdirected to that employee's VDU screen.

This chapter is an attempt to demonstrate the mathematical methods and costs of avoiding such embarrassments. As a more serious application, the authentication techniques described here are the mathematical basis for the treaty verification systems developed by Sandia National Laboratories to monitor compliance with comprehensive nuclear test ban treaties. For an account of these applications, we refer to Simmons (1983).

Signature schemes exist which are based on both conventional or *symmetric* cryptosystems and public-key or *asymmetric* cryptosystems. We deal with both kinds. First, however, we consider authentication.

12.2 Authentication in a communication system

We have already met a weak form of authentication when discussing the password problem in Section 10.1. Two users A (Alice) and B (Bob) of a system wish to communicate with each other. How does A convince B that it is A who is speaking, and vice versa?

One method of doing this is based on the same mathematical principle as public-key cryptosystems (PKCS); broadly speaking, any such system can be used in the following way as an authentication mechanism. Suppose that a PKCS has encrypting and decrypting functions e and d respectively. B can convince himself that he really is in contact with A by using the following protocol:

(I) B generates a 'random' item R which could have arisen as an output of the encrypting function e_A. That is,

$$R = e_A(x)$$

for some pseudorandom x.

(II) B sends R to A who then uses her own private key to decrypt R and then sends $d_A(R)$ back to B along the open line.

(III) B then checks that

(1)
$$e_A(d_A(R)) = R,$$

and accepts that A is the true correspondent only if (1) holds.

Example *Authentication using the Rabin system.* Suppose that we are operating a public-key system based on Rabin's method as described in Section 11.4. Without any real loss of generality, we may take the encryption function to be

$$e_A(x) = x^2 \pmod{N},$$

where N is the chosen public key of A. Then, if we wish to be sure we are in communication with A, we choose some large integer m (say) and send to A the 'random' item $R = m^2 \pmod{N}$. On receiving R, A proceeds to decode it by using her knowledge of factorization of N into primes p and q, and finds a square root m_1 of R. She returns m_1 to us.

Clearly we believe A is our true communicant if and only if the returned item m_1 satisfies $m_1^2 = R \pmod{N}$. □

Exercises 1. In an authentication system based on the RSA scheme, a user Alice has
12.2   chosen as her public key the integers $e = 7$ and $n = 77$. If she is sent by Bob the integer 23, how does she respond in order to authenticate herself?

2. In an authentication system based on the Rabin public-key system, Alice takes as her public key

$$B = 2, \qquad N = 209.$$

Bob sends her the integer $R = 168$. How does she respond in order to convince Bob they are in contact?

12.3   Signature schemes based on conventional cryptosystems

Instead of basing an authentication scheme on a public key system as described above, it is possible to use a symmetric cryptosystem, such as DES, to produce signed messages. We describe here two suggested approaches to the problem.

THE DIFFIE–LAMPORT SCHEME

This works as follows. The sender, Alice, who wishes to sign an n-bit binary message

$$M = M_1 \ldots M_n$$

chooses in advance $2n$ keys of the cryptosystem $\langle M, K, C \rangle$ being used. We denote them by

$$a_1, \ldots, a_n; \qquad b_1, \ldots, b_n.$$

These are kept secret.

If the encryption algorithm is e, Alice then generates $4n$ *validation parameters* $\{(X_i, Y_i, U_i, V_i) : 1 \le i \le n\}$, where the X_i and Y_i are in the domain of e and

(1) $$U_i = e(X_i, a_i) \quad \text{and} \quad V_i = e(Y_i, b_i) \quad (1 \le i \le n).$$

These validation parameters are sent in advance to the receiver Bob. They are also sent to an independent verifier such as a public registry.

Now suppose that Alice wishes to send a signed n-bit message $M = (M_1, \ldots, M_n)$. She adopts the following procedure: her signature will be the string

$$S = S_1 \ldots S_n,$$

where, for each i,

$$S_i = \begin{cases} a_i & \text{if } M_i = 0, \\ b_i & \text{if } M_i = 1. \end{cases}$$

The verification protocol of Bob is as follows: for each i $(1 \le i \le n)$, he uses the bit M_i and key S_i to check that

(2)
$$\begin{cases} \text{if } M_i = 0 \quad \text{then } e(X_i, S_i) = U_i, \\ \text{if } M_i = 1 \quad \text{then } e(Y_i, S_i) = V_i. \end{cases}$$

Bob accepts the signed message as genuine only if the verification procedure checks out for each i.

Although this system is simple to use and easy to understand, it has at least two obvious drawbacks. First, there has to be this precommunication of the validation parameters. More importantly, there is a mind-boggling expansion of the message by the signature. For example, with DES, where the keys are 64-bit strings, using the Diffie–Lamport scheme would involve a 64-fold expansion of the message.

RABIN'S PROBABILISTIC SIGNATURE SCHEME

An alternative approach to signatures was proposed by Rabin (1978). This works as follows. Let e be the encryption function of some standard cryptosystem $\langle M, K, C \rangle$ and let $(K_i : 1 \le i \le 2r)$ be a sequence of randomly chosen keys which the sender Alice keeps secret. The receiver Bob is then given a list of validation parameters (X_i, U_i) $(1 \le i \le 2r)$, where

(3)
$$e(X_i, K_i) = U_i \quad (1 \le i \le 2r),$$

and these parameters are again stored in some public record office.

Suppose now that Alice wishes to sign a message M. Her signature will be the concatenated string

$$S = S_1 S_2 S_3 \dots S_{2r},$$

where, for each i $(1 \le i \le 2r)$,

$$S_i = e(M, K_i).$$

Bob proceeds as follows: first he chooses, randomly or otherwise, r keys that he wishes to be disclosed. Let these be

$$K_{i_1}, K_{i_2}, \dots, K_{i_r}.$$

On being sent these keys by Alice, he checks that

(4)
$$e(M, K_{i_1}) = S_{i_1}, \qquad e(X_{i_1}, K_{i_1}) = U_{i_1},$$

and so on for $i_2, i_3, \dots, i_r$. He accepts the signature as being that of Alice only if all these checks hold. It is clear that the receiver's security depends on his belief that only a person holding the secret keys could have sent this signed message.

As for the sender, suppose that Alice wishes to challenge a message which she is alleged to have signed and which Bob has verified. Her protocol is clear: she must produce in front of a referee her secret keys

$$K_1, K_2, \ldots, K_{2r},$$

and the $2r$ validation checks

$$e(M, K_i) = S_i, \qquad e(X_i, K_i) = U_i$$

are made in public.

The ruling protocol of Rabin's system is that the challenge by the sender is upheld *only if* all the U_i but r or fewer of the S_i are correct.

Let us consider what this means in the three possible cases.

Case (a) fewer than r checks hold: then Bob should not have accepted (M, S) as a genuine signed message.

Case (b) exactly r checks hold: then, when choosing the r keys to be disclosed, Bob must have asked for precisely these keys. The probability of him picking this subset is given by

$$p_r = 1 \Big/ \binom{2r}{r},$$

and $p_r \sim 10^{-10}$ for $r = 18$.

Case (c) $r + 1$ or more checks hold: then the receiver (rightly or wrongly) is upheld.

Exercises 12.3

1. Using the Diffie–Lamport signature scheme based on the encryption

$$e(x, K) = x^K \pmod{N},$$

Alice chooses in advance the four keys

$$a_1 = 2, \quad a_2 = 5, \quad b_1 = 7, \quad b_2 = 3.$$

(a) If $N = 13$, what validation parameters does she send Bob?
(b) How would she sign the message $M = 10$?

2. In the above scheme, Bob receives the signed message $M_1 M_2 73$. If the message is genuine, what are M_1 and M_2?

12.4 Using public-key networks to send signed messages

The idea of using public-key systems to sign messages seems to be due to Diffie and Hellman (1976). They proposed a scheme whereby user A's signature for a message M should be a value S, depending on M and on information held secret by A, such that anyone else can verify the validity of A's signature S but no one else can forge A's signature on any messages.

One way in which this can be achieved is to use any public-key cryptosystem having the property that, for any user I, the encrypting and decrypting functions commute, that is:

(1)
$$e_I(d_I(x)) = x.$$

Given a system for which (1) holds, the following protocol allows A to send 'signed' messages to B, and these signatures are verifiable in the above sense.

We assume first, when discussing the protocol in outline, that the arguments of the various functions involved do lie in the domain of these functions. This is not always true in practice and can cause problems.

ENCODING ALGORITHM FOR ALICE

(I) Compute the *signature* S of the message M by using her own private key to obtain

$$S = d_A(M).$$

(II) Using Bob's public key, compute

$$C = e_B(S)$$

and send C to Bob.

What does B do? Well, on receiving C he can certainly decode it with his own private key to retrieve the signature S by the formula

$$S = d_B(C).$$

Now B can recover M from the signature S by applying A's public key to get

(2)
$$M = e_A(S)$$

because we have assumed the public-key system satisfies (1).

The receiver Bob is now in a very strong position. He has in his possession the pair (M, S). In case of dispute, if he needs to convince a judge that Alice did send the message, he asks her to produce the private key K_A used. Alice must produce her genuine private key, since it can be tested on the identity $d_A(e_A(x)) = x$, and then the judge need only verify that $d_A(M) = S$.

By the same token, this system keeps Bob honest. To see this, suppose that he altered the received message M to M'. Then, in order not to be exposed, he would also have to alter the signature S to $S' = d_A(M')$. But only Alice can do this.

Example *The RSA signature scheme.* We now examine in detail an implementation of the above signature protocol, using as our public-key system the RSA scheme. First note that, for this scheme, the encryption–decryption functions satisfy the necessary condition

(3)
$$e\bigl(d(x)\bigr) = x.$$

This is easy to prove: just follow through the proof given in Section 11.2 that $d\bigl(e(x)\bigr) = x$, and note that the d and e functions are of exactly symmetric form.

If we denote the public key of User I by (e_I, n_I) and the private key of I by d_I, then the encoding procedure for Alice who wishes to send Bob a signed message M is as follows. She first calculates

(4)
$$S = M^{d_A} \pmod{n_A},$$

and then encodes S by the cipher C where

$$C = S^{e_B} \pmod{n_B}.$$

On receiving C, Bob decodes C using

$$S = C^{d_B} \pmod{n_B}, \qquad M = S^{e_A} \pmod{n_A}.$$

He thus has the signed message (M, S) in his possession. □

There are various points to note. Apart from the fact that—for the scheme to work—(3) must hold, it must also be true that the signature S computed by A is within the range of the encryption procedure e_B.

This latter condition certainly may not hold when the system used is the RSA; the signature S may be a larger integer than the public key n_B. We can always avoid this by reblocking: in other words, readjusting the size of our message blocks so that they fall within the required range. However, a more elegant solution proposed by Rivest, Shamir, and Adleman (1978) is as follows:

Solution A threshold value h is adopted for the public-key system (say $h \sim 10^{199}$). Each user then maintains *two* public-key pairs, one for enciphering and one for signature verification. Denote these by (e_I, n_I) and (f_I, m_I) respectively, where I is the user in question. The rule adopted is that the encryption modulus n_I and signature modulus

m_I of each user I should satisfy

$$m_I < h < n_I.$$

Thus, if g_A is the private key of A corresponding to her signature pair (f_A, m_A), the modified protocol in the example above would be:
Calculate S by

(5)
$$S = M^{g_A} \pmod{m_A}.$$

Then, as before,

$$C = S^{e_B} \pmod{n_B}, \qquad S = C^{d_B} \pmod{n_B},$$

but M is recovered by

(6)
$$M = S^{f_A} \pmod{m_A}.$$

It is easy to verify that, for this system to 'work' and to allow messages to be signed and verified by all users of the system, all we need is that the set of messages M satisfies the condition

(7)
$$0 \le M \le \min m_I,$$

where the minimum on the right-hand side is over all users I of the system.

In practice, there are several drawbacks to the above signature system. To name but three:

(a) The sender Alice may deliberately 'lose' her private key, so that, unless it is deposited in a 'private key bank' before the start of the system, the messages sent by her could become unverifiable.

(b) The sender Alice may deliberately expose her private key d_A and, by so doing, allow all digital signatures attributed to her to become questionable.

(c) The time involved in encrypting, signing, decrypting, and verifying can be excessive; with information travelling as fast as it does nowadays (perhaps 1 million bits per second), there is much to be said for a faster signature scheme. This is the subject of the next section.

Exercises 12.4

1. In a signature scheme based on the RSA public-key system, the users Alice and Bob have public keys

$$e_A = 3, \quad n_A = 15, \quad e_B = 7, \quad n_B = 77,$$

respectively. Alice wishes to send the message $M = 4$ as a signed message to Bob. What integer will she send?

2. In the RSA signature scheme, prove that, if Alice signs message M_1 with S_1 and message M_2 with S_2, then she will sign the message $M_1 M_2$ with $S_1 S_2$, all arithmetic being modulo n_A.

12.5 Faster signatures but less privacy

If we are not too concerned with secrecy, then the encryption procedure described in the previous section can be modified so as almost to halve the processing time.

Consider the following algorithm.

ENCODING ALGORITHM FOR A

Compute the *signature* S of the message M by the rule

$$S = d_A(M)$$

and send S to B on an open channel.

The receiver B obtains M from S by using A's public key to get

(1)
$$M = e_A(S),$$

and then is in exactly the same position as regards proving that A actually sent the message as in the previous section. However, if an enemy should intercept the transmitted signature S *and* knows the source of the message (which is a very likely scenario), then the enemy can compute M by the same rule (1). Thus, this scheme has no real secrecy. However, it is much faster, and its speed makes it worthwhile for the many applications where the vital factor is authenticity rather than security. If we are prepared to forget about secrecy and concentrate on the verification problem, then the obvious protocol for A wishing to send a signed message to B is for her to send on an open line the signed message (M, S) where

$$S = d_A(M).$$

Example *The Rabin signature scheme.* Recall from Section 11.4 that the encryption function e of the public key system of Rabin is given by

$$e(x) = x(x + B) \pmod{N},$$

where N is the product of two primes (kept secret) and (N, B) is the public key of a typical user.

For simplicity, and without any real loss of generality, let us suppose that $B = 0$, so that

(2)
$$e(x) = x^2 \pmod{N}, \qquad d(x) = \sqrt{x} \pmod{N}.$$

Hence Alice will sign a message M by using the signature

$$S = \sqrt{M} \pmod{N_A};$$

here, in order to find $\sqrt{M}$, she uses her knowledge of the prime factorization pq of N_A exactly as in Section 11.4. The receiver of the signed message $(M, \sqrt{M})$ then only has to check its authenticity by looking up Alice's public key N_A and verifying that

$$(\sqrt{M})^2 = M \pmod{N_A}.$$

There is one fairly obvious snag in the procedure described: not all integers less than N have square roots modulo N. (Recall from Section 11.4 that M has such a square root only when it is a quadratic residue.) To avoid this, Rabin introduces the following modification of the basic idea described above.

By convention, when wishing to sign a given message M, Alice adds as a suffix a binary string U of agreed length k. The choice of U is randomized each time a message is to be signed. Alice now compresses the concatenated $M_1 = MU$ by a hashing or compression function c so that

(3)
$$c(MU) = c$$

is an integer $\leq N_A$. (The function c is publicly known and therefore publicly verifiable.)

Alice now uses her private key to check whether this integer c has a square root modulo N_A. If not, she chooses (again at random) another k-bit suffix U, and repeats the above until she eventually finds a U such that the corresponding c defined by (3) has a square root S.

It can be shown by number-theoretic arguments that on average the number of trials needed to produce an integer c of the form (3) that has a square root modulo N_A is low.

Alice now uses as her signature on the message M the pair (U, S). Anyone wishing to verify that Alice did indeed send the message M has only to

(a) compute $c(MU) = M_1$,
(b) verify that $S^2 = M_1 \pmod{N_A}$. ☐

Exercises 12.5
1. In the signature scheme based on Rabin's method (2), Alice has public key $N_A = 35$ and wishes to sign the messages $M_1 = 9$ and $M_2 = 29$. What are the respective signatures S_1 and S_2?
2. If a user of a Rabin-type signature scheme signs the message M with S, how would the same user sign the message $4M$?

12.6   Attacks and cracks in trapdoor signature schemes

The fundamental characteristics of the signature schemes we have discussed have been (a) that each message M is accompanied by a signature S and (b) that underpinning the scheme is a 'trapdoor function' f such that

$$f(S) = M$$

but only the sender A possesses the 'trapdoor' information to invert the function to obtain the signature.

We now follow Goldwasser, Micali, and Rivest (1984) and try to characterize some of the possible attacks on such a system under the following headings.

Direct attack

This is an attack in which the only information available is A's public key.

Known-message attack

The enemy has a set of signatures $S_1, \ldots, S_k$ for a known set of messages $M_1, \ldots, M_k$.

Generic chosen-message attack

The enemy can choose messages $M_1, \ldots, M_k$ for the sender to sign, but has to make this choice before knowing A's public key. For example, they could be k messages chosen at random.

Directed chosen-message attack

The enemy is allowed to choose which messages he would like signed after seeing the sender's public key.

Adaptive chosen-message attack

In this case, the enemy can decide which message he would like A to sign after seeing some previously signed messages. In other words, his choice of message M_i depends on $(M_1, S_1), \ldots, (M_{i-1}, S_{i-1})$.

Obviously these attacks are ordered in increasing order of their power.

In the same way that there exists a variety of attacks, there also exists a variety of ways in which a signature system may be compromised. We list some of these in decreasing order of their severity.

Total break

The sender's secret trapdoor information (i.e. the private key) is obtained.

Universal forgery

The enemy is able to find an efficient signing algorithm having exactly the same effect as A's signing algorithm, for example, computing the RSA function without factoring the public key.

Selective forgery

The enemy may be able to forge a signature for a specific message chosen a priori.

Existential forgery

The enemy can forge a signature for at least one message (the enemy has no control over which message is to be signed).

Note   We also demand that, when we speak of a scheme being broken in any of the above senses, then we insist that it be broken with non-negligible probability, that is, for at least some positive fraction of all the possible keys.

We now give some examples to show the extent to which our previously discussed schemes are vulnerable to various attacks.

Example 1   The RSA signature scheme is selectively forgeable using a directed chosen-message attack.

Proof   If (M_1, S_1) and (M_2, S_2) are two signed messages from the same person, then it is not hard to check that the signature of the product $M_1 M_2$ is the product $S_1 S_2$.   □

Example 2   It is possible to break the Rabin scheme by a directed chosen-message attack.

Proof   Suppose that we get A to sign a message

$$M = x^2 \ (\mathrm{mod}\ N),$$

where x was picked at random and N is the public key of A. Then, with probability $\frac{1}{2}$, A is going to sign M with 'another' square root of M—say y—so that

$$x^2 = M \ (\mathrm{mod}\ N), \qquad y^2 = M \ (\mathrm{mod}\ N),$$

giving

$$x^2 - y^2 = 0 \pmod{N}.$$

But $x - y \neq 0 \pmod{N}$ and $x + y \neq 0 \pmod{N}$, and so we have

(1) $\gcd(x + y, N)$ is a prime factor of N.   □

A more general result is

(2)   Any trapdoor signature scheme is existentially forgeable by a direct attack.

Proof   A valid signed message (M_A, S_A) can be obtained by taking any possible signature S_A and using the public verification $e_A(S)$ to get a signed message $(e_A(S), S)$. A possible heuristic for dealing with this problem is to require that the message space be sparse, so that it is highly unlikely that $e_A(S)$ belongs to it.   □

To get around these and other difficulties, Goldwasser, Micali, and Yao (1984) have recently introduced the notion of a *strong signature scheme*. We will not go into the details here; it is closely related to the probabilistic encryption methods discussed in the next chapter.

PROBLEMS 12

1. In a signature scheme based on the RSA public-key system, user A has public key $e = 11$, $n = 899$. How will she sign the message 876?
2. Using the Rabin probabilistic signature scheme with encryption function

$$e(x, K) = x^K \pmod{13},$$

Alice sends Bob the four validation parameters

$$(2, 10), \quad (11, 7), \quad (7, 8), \quad (6, 3).$$

How will she sign the message $M = 8$?
3. In Problem 2, why would there be difficulties if Alice sent the validation parameter $(3, 1)$ instead of $(2, 10)$?
4. As a slightly more realistic example of the RSA problem, find the private key of a user whose public key is

$$n = 4058287248123404834599,$$

$$e = 1897225149044257283231.$$

(This is the example used to illustrate the authentication scheme in use at Sandia and is taken from Simmons (1983).)
5. Suppose that we know the signatures S_1 and S_2 of messages M_1 and M_2

using the Rabin scheme

$$E(x) = x(x + B) \pmod{N}.$$

Show that, for certain values of the public key B, it is easy to sign the message $M_1 M_2$.

6. Show how the public-key system based on discrete logarithm and introduced in Section 11.5 can be used to give a signature scheme. Verify that it is vulnerable to existential forgery.

(Elgamal, 1985)

13

Randomized encryption

13.1 Introduction

In this final chapter we shall describe a selection of the more exciting recent developments in the area of codes and cryptography, which involve randomization in the encryption process.

The first problem discussed is that of finding encryption schemes that are *semantically secure*. These are schemes that ensure the secrecy of all partial information about transmitted messages, and an example of such a cryptosystem due to Goldwasser and Micali (1982) is described in Section 2. This is closely related to the unresolved problems mentioned earlier of finding 'cryptographically safe' pseudorandom numbers. These are sequences in which no adversary can, in a reasonable time, predict with better than average success the next integer to be emitted.

The same theme of randomization in the encryption process underlies a practical solution to a fundamental problem solved by A. D. Wyner in 1975 when he brought together information theory, codes, and cryptography in his study of what is now known as Wyner's wiretap channel. Before discussing these problems, however, we introduce the basic idea.

HOMOPHONIC SUBSTITUTION

The simplest randomized encryption scheme is what is known as a *homophonic substitution cipher*. The message M consisting of a string from an alphabet Σ_1 is encrypted into a random cipher text C of symbols from an alphabet Σ_2, where $|\Sigma_2| \geq |\Sigma_1|$.

If Σ_1 has t symbols, then Σ_2 is partitioned into disjoint nonempty subsets A_i $(1 \leq i \leq t)$ so that symbol s_j from Σ_1 is encrypted into a random element of the subset A_j. The key K of the system is the ordered partition $(A_1, \ldots, A_t)$. $\square$

The way in which randomization obtains higher levels of security is clear. It increases the number of keys, thus increasing the equivocation. The cost of a randomized encryption scheme is that, since for each key each message corresponds to several ciphertexts, the ciphertext space will be larger than the message space. To accomodate this, there must be a certain amount of expansion in the communication channel, since more information is needed to specify the ciphertext than is needed to identify the message. This is called *bandwidth expansion* and appears to be unavoidable in any randomized encryption scheme.

This idea is not new. According to Simmons (1985b), Gauss believed he had discovered an unbreakable cipher by introducing homophones. A later example of a homophonic cipher was the Jefferson–Bazeries wheel cipher and its variations used by the USA in both world wars. Other examples and more recent randomization techniques are described by Rivest and Sherman (1983).

Exercise 13.1

1. If a homophonic substitution scheme is based on a mapping of the binary alphabet into the 26-letter Roman alphabet, how many keys does it have?

13.2 Semantic security and the Goldwasser–Micali scheme

Encryption schemes that ensure the secrecy of all partial information about messages are obviously important goals in cryptography. Such a scheme is described as *semantically secure.*

We have not addressed the problem of semantic security earlier. For example, the sort of security measures we have been discussing do not rule out the possibility that perhaps even as much as every other bit of a string encrypted by the RSA method could be easily computed.

In this chapter we shall indicate some of the progress—most of it very recent—which has been made towards the resolution of such problems. Because a full rigorous description is difficult and demands more machinery than is appropriate at this level, the treatment will be fairly informal but hopefully it will give some insight into the problems and ideas circulating in this field.

Example

Consider a public-key scheme based on the difficulty of factoring an integer N that is the product of two primes p and q. The general problem of determining p and q from N is probably very difficult; however, when the least-significant digit of N is 3, it is easy to obtain the partial information that the least-significant digits of p and q are either 1 and 3 or 7 and 9.   □

A particular case of semantic security is *bit security,* which is concerned with ensuring not only that the whole message is not recoverable but also that individual bits of the message are not obtainable. Interest in these problems of semantic and bit security seems to have been sparked off by the work of Goldwasser and Micali (1982) which first raised these questions, and answered them in one respect, by proposing a new semantically secure scheme based on probabilistic encryption.

The scheme is an extension of the public-key idea, but encrypts messages bit by bit. The actual encryption of a single bit depends on the message and on the result of a sequence of coin tosses. It turns out to be completely secure with respect to the notions of bit and semantic security, provided that the number-theoretic problem of deciding quadratic residues is 'hard'.

Before describing the concrete implementation of the scheme, we introduce the quadratic-residue problem.

QUADRATIC RESIDUES

We have already met the problem of deciding whether there exists a solution x to the congruence

(1)
$$x^2 = b \pmod{N}$$

in our discussion of Rabin's public key-scheme in Section 11.4.

If (1) has a solution, then b is called a *quadratic residue* modulo N.

The Goldwasser–Micali scheme (GM-scheme) is based on the belief that, for certain N and b, deciding whether b is a quadratic residue modulo N is a 'hard' problem. It is certainly a well-known hard problem in number theory, being one of the four algorithmic problems discussed by Gauss in his *Disquisitiones Arithmeticae* in 1801. Basic properties of quadratic residues are the following statements whose proofs can be found in any standard text on number theory.

(2)   For any odd prime p, exactly half the positive integers less than p are quadratic residues modulo p.

(3)   (Euler's criterion) If a is coprime with the odd prime p, then a is a quadratic residue modulo p if and only if

$$a^{\frac{1}{2}(p-1)} = 1 \pmod{p}.$$

(4)   If a is an integer in the range $\{1, \ldots, N\}$ and N is the product of the two primes p and q, then a is a quadratic residue modulo N if and only if it is a quadratic residue with respect to both p and q.

We now describe the actual encryption procedure.

It works by taking a message M as a sequence $M_1 \ldots M_n$ of bits and encrypting each bit M_i as a string.

Suppose that Alice wishes to send M to Bob. First Bob has to set up his public/private keys by the following procedure.

(I) Randomly select two k-bit primes p and q and let $N = pq$. The factorization (p, q) of N will be the receiver's *private key*.

(II) Choose an integer y at random from the set of integers coprime with N such that y is a quadratic nonresidue with respect to N.

(III) Publish as his *public key* the pair (N, y).

The sender A now encrypts her message $M = M_1 M_2 \ldots M_n$ by the following encryption algorithm.

Encryption Algorithm

For each bit M_i, pick x at random from the set of integers less than and coprime with N. Encrypt M_i by

$$e(M_i) = \begin{cases} yx^2 \ (\mathrm{mod}\ N) & \text{if } M_i = 1, \\ x^2 \ (\mathrm{mod}\ N) & \text{if } M_i = 0. \end{cases}$$

Thus the final cryptogram C corresponding to the message M is

(5)
$$C = e(M_1)e(M_2) \ldots e(M_n).$$

Since N is a $2k$-bit integer, it is clear that the cryptogram C is a much longer string than M.

We will assume for the moment that the above encryption and setting up of the public-key register is achievable in polynomial time (in fact in time $O(nk^2)$).

Let us now consider the decrypting problem facing Bob.

Decryption Algorithm

Suppose we denote the integers $e(M_i)$ by e_i ($1 \leq i \leq n$). Then the procedure for decoding is easy:

if e_i is a quadratic residue modulo N, then $M_i = 0$

if e_i is not a quadratic residue modulo N, then $M_i = 1$.

The trapdoor information of Bob is that, because he knows the factorization of N into primes p and q, he can decide very quickly by (3) whether or not e_i is a quadratic residue modulo p and modulo q. Using (4), this tells him whether the received e_i is a quadratic residue modulo N, and hence decryption is easy.

Example Suppose we consider the GM-scheme with security parameter k. The encryption algorithm e will send the bit $M_i = 1$ into one of

$$yx_1^2, yx_2^2, \ldots, yx_{\phi(N)}^2 \quad (\mathrm{mod}\ N),$$

where $x_1, \ldots, x_{\phi(N)}$ are the integers coprime with and less than N and (N, y) is the receiver's public key. It will send the bit $M_j = 0$ into one of

$$x_1^2, x_2^2, \ldots, x_{\phi(N)}^2 \quad (\text{mod } N).$$

In order to be safe, N must be large enough for it to be impractical to decide quadratic residues. Since the decryption algorithm relies on being able to factor N, this means N must be at least 200 digits long. Hence it is manifestly inefficient and is going to slow down the rate of communication drastically.   □

We now describe the sort of enemy we are prepared to encounter. We assume that we are dealing with a *passive line tapper* who knows the message space, the probability distribution of messages, and the encryption algorithm, and is given the ciphertext and tries to retrieve the message.

If we allow the line tapper unlimited time and computing powers, he will be able to decipher the message just by using exhaustive search. What we really want is a situation in which the adversary cannot hope to decipher the message in any reasonable span of time.

If we consider the GM-scheme, we note two distinct features:
(a) Each *bit* is encrypted as an integer modulo N and hence is transformed into a 2k-bit string.
(b) The encryption function is random in that the same bit is transformed into different strings depending on the choice of the random quantity x. For this reason it is called *probabilistic encryption*.

The security of the system depends on the supposed inability of the line tapper to decide quadratic residues modulo N, and this in turn depends on the length of N, namely $2k$. Accordingly we can regard k as the *security parameter* of the system.

Once k is decided as our parameter, what we now ask is that nobody with a time resource bounded by a polynomial in k can 'crack the system'. What Goldwasser and Micali prove is that their system has this property in an extraordinarily strong sense. We describe their results informally.

Suppose that we think of the enemy as a seller of code-cracking equipment. In order to demonstrate his wares, he sets up his polynomially bounded line tapper T. He then employs a *message finder F*, also with computing resources bounded by a polynomial in k, to search the message space to find two messages M_1 and M_2 such that the encryptions of M_1 and M_2 are distinguishable by T.

We say a system is *polynomially secure* if a polynomially bounded message finder F cannot find two messages M_1 and M_2 whose encryptions are distinguishable by a polynomially bounded line tapper T. That is, given C (an encryption of either M_1 or M_2), T should have no advantage in understanding which of the two messages M_1 and M_2 is being encoded by C.

What Goldwasser and Micali prove about their encryption scheme is that it is semantically secure against any threat from a polynomially bounded attacker, provided that the quadratic-residue problem is 'hard' in the following sense. For k a positive integer, the set H_k of *hard composite integers* is defined by

$$H_k = \{N : N = pq \text{ where } p \text{ and } q \text{ are } k\text{-bit primes}\}.$$

THE ASSUMPTION THAT THE QUADRATIC-RESIDUE PROBLEM IS INTRACTABLE

Let P and Q be fixed polynomials. For each integer k, let C_k be a Boolean circuit, with two $2k$-bit inputs and one Boolean output, that has the following properties.
(a) Given input (n, x), then, for a fraction $1/P(k)$ of $n \in H_k$, it correctly decides whether or not x is a quadratic residue modulo n for each integer x less than and coprime with n;
(b) of all such circuits satisfying (a) it has a minimum number of gates.
Then, for all sufficiently large k, the size—that is number of gates in the circuit C_k—is at least $Q(k)$. $\qquad\square$

In the light of the results obtained by Goldwasser and Micali, an obvious question to ask is: how secure are the bits of more practical schemes such as the RSA and Rabin public key schemes? This is still very much an area of ongoing research; we present a brief resumé.

Goldwasser, Micali, and Tong in 1982 proved that specific bits of both the RSA and Rabin scheme were secure in the sense that finding the least significant bit of the RSA cleartext is equivalent to breaking the RSA system. This result was improved by Ben-Or, Chor, and Shamir (1983), in which they show that any method that determines the least-significant bit must either have an error probability of $\frac{1}{4} - \varepsilon$, or be able to break the RSA encryption completely. This was improved by Alexi, Chor, Goldreich, and Schnorr (1984), who strengthened the above result by replacing $\frac{1}{4} - \varepsilon$ by $\frac{1}{2} - 1/(\log N)^c$ where N is the modulus of the encryption scheme. In other words, they show that getting 'any advantage' in computing the least significant bit is as hard as deciphering the message.

            2. In the GM-scheme, a receiver A has public key $N = 35$, $y = 3$. How would
               she decode the cipher text $(29, 13)$?

13.3 Cryptographically secure pseudorandom numbers

The problem of finding 'unpredictable' pseudorandom numbers is
intimately related to randomized encryption. Given a method of
generating such sequences, it can be used as the source of
approximate one-time pads that are safe in the way that the linear
shift-register sequences described earlier are not.

A secure probabilistic encryption scheme, based on the ideas of
Goldwasser and Micali but much more efficient to implement, has
been announced recently by Blum and Goldwasser (1985). The basic
idea of the scheme is to encrypt the message by taking its modulo 2
sum with the output of an unpredictable pseudorandom sequence
generator.

Suppose A and B wish to communicate secretly (and therefore
share) a common random-number generator and a common secret
seed s. In order to send a block of plaintext $M_1 M_2 \ldots$ to B, sender A
uses s to generate a sequence of pseudorandom numbers
$X_1, X_2, \ldots$, and forms the cryptogram $C = C_1 C_2 \ldots$, where

$$C_i = M_i + X_i \pmod{2}.$$

A danger to the system occurs when an enemy has some side
information enabling him to find out some of the X_k. Based on these
values, he may be able to generate the rest of the sequence and thus
break the cryptogram.

Example   One of the most popular and fast methods of obtaining pseudorandom sequences is to use the method of linear congruential generators. These consist of a *modulus N*, a *multiplier a* relatively prime to
N, and an *increment c*. Starting from a random *seed* X_1, the sequence
$(X_n : 1 \leq n < \infty)$ is given by

(1) $$X_{n+1} = aX_n + c \pmod{N}.$$

Thus the X_i are all integers between 0 and $N - 1$. The sequences
produced have been shown to satisfy many and various statistical
tests of randomness for proper choices of the parameters a, c, and N
(for details see Knuth, 1969). Because of this and their ease of
generation, this *linear congruential* method is widely used.

However it is cryptographically insecure. Plumstead (1982) has shown how to recover the parameters a, c, and N from a knowledge of all the bits of several consecutive X_k. More recently Frieze, Kannan, and Lagarias (1984) have shown that the rest of the sequence is almost always predictable in polynomial time given only at least $\frac{2}{3}$ of the leading bits of the first few numbers generated. □

An informal description of a *cryptographically secure* random number generator (RNG) is as follows. Starting with a *seed s* of (say) n true random bits, the generator must use it to produce, in time polynomial in n, a sequence of bits

$$X_1, X_2, \ldots, X_{n^k}$$

with the property that, given the generator and the first t of the X_i but *not* the seed s, it is computationally infeasible to predict X_{t+1} with better than probability $\frac{1}{2}$ of success. Such a generator is said to *pass the next-bit test*.

In 1982, Blum and Micali produced a generator that was cryptographically secure in the above sense, *provided* that the discrete-logarithm problem was 'hard' in the well-defined sense that, once the inputs were sufficiently large, any efficient procedure for calculating the discrete logarithm would fail for at least some fixed fraction of the inputs. Although the Blum–Micali generator is a landmark in the history of the problem, it is not easy to use or describe and, in a sense, has been overtaken by more practical and efficient systems. One of these is what is known as the $x^2 \bmod N$ *generator,* which Blum, Blum, and Shub (1983) proved to be cryptographically secure with respect to next-bit tests, provided that any efficient procedure for solving the quadratic residue problem will fail for some fixed positive fraction of the inputs. This generator works as follows.

Take N to be the product of two equal length primes p and q, each congruent to 3 modulo 4, and take Y_0 to be any quadratic residue modulo N.

Then define the pseudorandom sequence $(X_i : 0 \le i < \infty)$ by

$$X_i = \text{parity } (Y_i),$$

where

$$Y_{i+1} = Y_i^2 \pmod{N} \quad (0 \le i < \infty).$$

An objection to the notion of regarding generators passing all next-bit tests as cryptographically secure is that the cryptanalyst may use a different approach to analyse the generator. For example, he or she may work backwards and examine the last few bits in an attempt to predict earlier bits. This problem was addressed by Yao in 1982 in a fundamental paper in which he introduced the much stronger notion of a *polynomial-time statistical test*. This can be formally defined as follows.

Let G be a pseudorandom number generator which stretches a k-bit random seed s into a $p(k)$-bit sequence, where p is some fixed polynomial. Let S_k be the collection of $p(k)$-bit sequences produced by G from the k-bit seeds.

A *polynomial time statistical test* is a randomized algorithm $\mathcal{A}$ which, given an input string of length n, stops in time polynomial in n and outputs 0 or 1. The generator G *passes* the test $\mathcal{A}$ if, for any integer t and all sufficiently large k, we have

$$|P_k^S - P_k^R| < k^{-t},$$

where P_k^S is the probability $\mathcal{A}$ outputs 1 on a randomly chosen string from S_k and P_k^R is the probability $\mathcal{A}$ outputs 1 on a truly random string of the same length.

We say a generator G is *cryptographically secure* if it passes all polynomial-time statistical tests. In other words, a pseudorandom number generator G is cryptographically secure if there is no effective algorithm which, in polynomial time, will separate the output of G from that produced by a truly random source.

A major result of Yao is the following. We slightly tighten the definition of a *next-bit test* by regarding it as a randomized polynomial algorithm $\mathcal{T}$ which, on being presented with input k and the first i bits of a string s randomly chosen from S_k, gives as output a bit b. Let p_k^s be the probability that the bit b equals the $(i+1)$th bit of s. We say G *passes* the next-bit test $\mathcal{T}$ if, for any $t > 0$ and sufficiently large k,

(2)
$$|p_k^s - \tfrac{1}{2}| < k^{-t}.$$

Theorem *A pseudorandom number generator passes all polynomial-time statis-*
(Yao) *tical tests if and only if it passes all polynomial time next-bit tests.*

Proving this in one direction is straightforward (and intuitively obvious). Proving that the ability to pass all next-bit tests implies the ability to pass all statistical tests is far from obvious. For a proof, see Goldreich, Goldwasser, and Micali (1984).

The question remains: do there exist generators that pass all polynomial-time statistical tests? Such generators are what we now regard as *cryptographically secure*.

By Yao's theorem, it is sufficient that they pass all polynomial-time next-bit tests. Over the last few years, several authors have proposed possible secure generators; in each case, the security of the generator is, like those discussed earlier, dependent on the intractability of various number-theoretic problems. At the time of writing, that proposed by Alexi, Chor, Goldreich, and Schnorr (1984) would seem to be both the most efficient and—being based on the intractability of factoring—also the most secure.

We close this brief introduction to this rather difficult area by returning to the topic of one-way functions. At the intuitive level at least, a cryptographically secure random-number generator corresponds almost exactly with our notion of a *one-way* function as discussed in Chapter 10. Given the output of such a generator, it is computationally infeasible to find out anything about the seed in any reasonable (polynomial) amount of time.

Exercises 13.3

1. Prove that, in the linear congruential method of generating pseudorandom numbers, if the modulus N is leaked to an enemy, then the output X_k, X_{k+1}, X_{k+2} is insecure for any integer k, since from this information it is possible to infer the values of the multiplier a and the increment c.

2. Suppose the congruence

$$X_{n+1} = aX_n \pmod{N},$$

with X_0 coprime to N, is used as a generator of pseudo-random numbers, Prove that, when N is a prime p, the maximum period is $p - 1$ and that, for suitable choice of a, this period is attainable.

13.4 Wyner's wiretap channel

It is appropriate to round off our discussion of randomized encryption by describing briefly a fundamental paper of A. D. Wyner, which brings together the problems of cryptography, coding, and information theory in an attractive and practical way. The basic problem studied by Wyner is the following. A source $\mathscr{S}$ emits information at constant rate, and this has to be transmitted as reliably as possible to a legitimate receiver while keeping the wiretapper (enemy) as ignorant as possible. The situation is as represented in Fig. 1 and for simplicity, we will assume $\mathscr{S}$ is a memoryless source.

We first consider the special case where the main channel is a noiseless memoryless binary channel, and the wiretap channel is a binary symmetric channel with bit error probability $p < 1$.

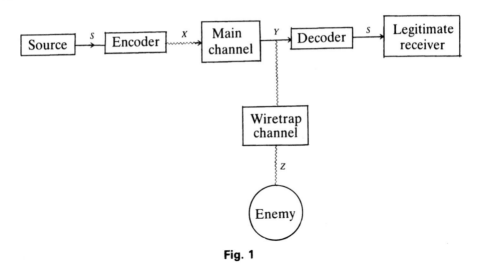

Fig. 1

A first solution to this problem is very simple:

Encryption Algorithm
Encrypt 0 as a random binary string of length N containing an even number of 1's.
Encrypt 1 as a random binary string of length N containing an odd number of 1's.

Clearly decryption is easy, the receiver has only to count the parity of the number of 1's in any block of length N. Thus there is no space or storage requirement.

The enemy, on the other hand, is baffled. To see why, recall that $p < 1$ is the probability that a bit is wrongly transmitted down the wiretap channel. Then the probability it is correctly received by the wiretapper is given by P_c, where

$$P_c = P \text{ [there are an even number of errors in } N \text{ symbols]}$$
$$= q^N + \binom{N}{2}q^{N-2}p^2 + \ldots \qquad (q = 1 - p)$$
$$= \tfrac{1}{2}[(q + p)^N + (q - p)^N] = \tfrac{1}{2}[1 + (1 - 2p)^N].$$

Thus $P_c \approx \tfrac{1}{2}$ for large N. Hence, whatever binary message is transmitted, the enemy wiretapper will be getting an almost completely scrambled text. However, there is a big price to be paid, namely the rate of transmission has been reduced by a factor of $1/N$.

Suppose instead we break the source stream up into blocks S of length k, and use an error-correcting code to encode each such block by the n-bit X so that the *rate R* of the code is k/n. If Z is the n-bit sequence received by the wiretapper, we let $H(X \mid Z)$ denote the wiretapper's equivocation so that

$$d = H(X \mid Z)/k$$

represents the equivocation of the wiretapper per bit transmitted.

Clearly, the parameters R and d are pulling against each other in that, as R increases, d decreases, and vice versa. What Wyner was able to prove was the following striking analogue of Shannon's noisy coding theorem. Let us call an (R, d) pair *achievable* if, for all $\varepsilon > 0$, there exists an encoder–decoder as shown, with parameters n and k such that

$$k/n \geq R - \varepsilon, \qquad k^{-1}H(X \mid Z) \geq d - \varepsilon, \qquad P_e \leq \varepsilon,$$

where P_e denotes the decoded-bit error rate at the legitimate receiver.

Theorem 1 *With the main channel noiseless and the wiretap channel binary symmetric with error probability p, a pair (R, d), with $0 < R, d < 1$, is achievable if and only if*

$$Rd \leq -p \log p - (1 - p)\log(1 - p).$$

In order to describe Wyner's more general result, consider what is being aimed at: a maximization of the rate of reliable communication to the legitimate receiver but at the same time keeping the uncertainty of the wiretapper above some threshold.

In terms of entropy, the ideal would be to maximize

$$I(X \mid Y) = H(X) - H(X \mid Y)$$

and simultaneously to minimize the information

$$I(X \mid Z) = H(X) - H(X \mid Z).$$

For any rate $R \geq 0$ define $\rho(R)$ to be the set of those probability distributions p on the input X to the main channel such that

$$I(X \mid Y) \geq R.$$

Let C_M be the capacity of the main channel, and let

$$\Gamma(R) = \sup_{p \in \rho(R)} [I(X \mid Y) - I(X \mid Z)].$$

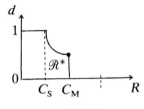

Fig. 2

Theorem 2   *The set of all achievable (R, d) pairs is given by*

$$\mathscr{R}^* = \{(R, d) : 0 \le R \le C_M, 0 \le d \le 1, Rd \le \Gamma(R)\}.$$

We illustrate what this theorem is saying in Fig. 2.

Here, the quantity C_S indicating the maximum value of R for which $d = 1$ represents the *secret capacity* of the system. At any rate below C_S, the transmitter may communicate with the legitimate receiver while keeping the uncertainty of the wiretapper at a maximum.

We close by indicating how randomized encryption leads to a practical solution of the wiretap problem in the particular case that the main channel is noiseless. For the moment, fix N and let $\mathscr{C}$ be a 'good' linear code with 2^{N-k} codewords of length N. Partition the set V_N of all binary N-vectors into the 2^k cosets of $\mathscr{C}$ in V_N.

There are 2^k possible messages of length k; also, if M_i is the ith message and C_i the ith coset, then the encryption function e is defined by

$$e(M_i) = X,$$

where X is a random member of coset C_i. By taking N sufficiently large and using Elias' Theorem 4.9.1, which assures the existence of linear codes arbitrarily close to channel capacity, this gives an encryption algorithm and 'constructive proof' of the theorem for this special case. The general case is complicated, and for it we refer to Wyner's original paper.

Exercises 13.4   1. Show that $\Gamma(R)$ is nonincreasing in R and satisfies

$$C_M \ge \Gamma(R) \ge C_M - C_{MW},$$

where C_{MW} is the capacity of the concatenation of the main channel with the wiretap channel.

2. Show that, when the main channel is noiseless and the wiretap channel is binary symmetric with bit error probability p, then

$$\Gamma(R) = -p \log p - q \log q \quad (0 \le R \le 1),$$

where $q = 1 - p$. What is the secret capacity in this case?

3. Verify that, when the wiretap channel is a perfect scrambler, the secret capacity equals the capacity of the main channel.

13.5 Effective entropy

We close this introductory text in the same way as we started it, by discussing the concepts of entropy and information.

In his 1982 paper referred to in Section 3, A. Yao also introduced a new measure of uncertainty and information based on computational complexity and which he called *effective entropy*. Many of the proofs of results announced by Yao have still not been published, and the underlying ideas are very much an area of ongoing research. Because of this, and the difficulty of some of the ideas, we shall be highly discursive. The aim is just to plant the seed of Yao's ideas before a wider audience, by getting them to consider the following examples.

Example 1 Let Σ be the collection of all k-bit binary strings ($k \sim 10^4$). For any x and m that are integers of 100 bits, define the string

$$\alpha_{x,m} = c_1 c_2 \ldots c_k,$$

where

$$c_j = \text{parity} \, (\text{mod}_m \, x^j),$$

$\text{mod}_m \, x^j$ denoting the unique integer in $\{0, \ldots, m-1\}$ congruent to x^j modulo m. Let Λ be the subset of Σ consisting of all strings $\alpha_{x,m}$. Thus

$$|\Lambda| = 2^{200}, \qquad |\Sigma| = 2^{10000}.$$

Consider a source $\mathcal{S}$ which has alphabet Σ but which emits only elements of Λ and those independently and with equal probability. Then the entropy

$$H(\mathcal{S}) \leq 200.$$

Hence by the noiseless coding theorem of Chapter 2, a source message of length n has an encoding of average length $200n$ bits. In fact, the sender Alice could use the representation

$$\alpha_{x,m} \mapsto (x, m)$$

which represents $\alpha_{x,m}$ by a 200-bit string. However, in order to do this, Alice has to obtain x and m from the 10 000-bit string $\alpha_{x,m}$, since this is all she knows. To quote Yao, 'It is not obvious that this can be done in a reasonable amount of computing time'. $\square$

Example 2   Suppose a message M is being sent over a public-key system by the RSA method using some large modulus N. Suppose also that the ciphertext C is intercepted (no big deal since it has been transmitted over a public network). Then we have the conditional entropy

$$H(M \mid C) = 0.$$

Thus there is no uncertainty about M and yet, for large N, there is certainly an 'effective uncertainty', namely one induced by the cryptanalyst's inability to factorize the public key N quickly.   □

The above examples hopefully give some idea of what 'effective entropy' means. Producing formal mathematical definitions and proofs of theorems seems to be difficult.

PROBLEMS 13

1. A homophonic substitution scheme maps an alphabet $\Sigma \mapsto \Sigma'$ in such a way that each symbol of Σ is mapped into an element of an m-element subset of Σ'. If $|\Sigma'| = N = m \, |\Sigma|$ and $|\Sigma| = t$, find the number of distinct keys.

2. Suppose that an encryption scheme encodes an integer x by taking a large known prime p and secret key $a < p$, where a is primitive modulo p, and

$$C = e(x) = a^x \pmod p.$$

Show that, given C, an enemy can find in polynomial time whether the least significant binary digit of x is 0 or 1.

3. (i) Show that, if a and b are both quadratic residues modulo a prime p, or if neither are, then ab is also a quadratic residue modulo p.

   (ii) Show that, if exactly one of a and b is a quadratic residue modulo p, then ab is not.

   (iii) Show that, if $c = d \pmod p$, then c is a residue modulo p if and only if d is.

4. Show that the assumption that the quadratic-residue problem is intractable, as used in the Goldwasser–Micali scheme, is at least as strong as the demand that there is no randomized polynomial-time algorithm for deciding quadratic residues.

5. A receiver in the Goldwasser–Micali scheme has a public key $N = 77$. In how many distinct ways could the message 101 addressed to this receiver be encrypted?

6. The encryption procedure for the Goldwasser–Micali scheme involves picking a 'random' y that is a quadratic nonresidue with respect to the chosen primes p and q. Show tht there is a randomized algorithm for picking such a y in time $O(\log N)^2$, where $N = pq$.

7. Consider a pseudorandom number generator used to produce a sequence $X_0, X_1, \ldots$ of integers in the range $\{0, \ldots, N\}$ by a rule of type

$$X_{i+1} = f(X_i) \quad (0 \le i < \infty).$$

Show that any such system must be ultimately periodic in the sense that there exist integers c and d such that $X_{n+d} = X_n$ for all $n \ge c$. (The least positive integer for which this is true is the *period* of the sequence.)

8. The $1/P$ *generator* (*base* 10) is a pseudorandom sequence generator which, with seed $X_0 = r$ and with P a prime such that 10 is a primitive root of P, produces the sequence of digits $X_1 X_2 \ldots$ which is the decimal expansion when r ($<P$) is divided by P. For example, with $P = 7$, we get the sequence

$$142857142 \ldots \text{ since } \tfrac{1}{7} = 0.1428571 \ldots$$

(a) Show that the sequence produced must be ultimately periodic.
(b) Show that the period is $\le P - 1$.
Note: Blum, Blum, and Shub (1983) show that sequences produced by this method are *not* cryptographically secure: it is possible to find P from any $2|P| + 1$ successive elements of the sequence, and this leads to an efficient test for separating sequences produced by this method from purely random sequences.

9. Show that a pseudorandom sequence $(X_0, X_1, \ldots)$ generated by a recurrence relation of the type

$$X_{n+1} = f(X_n, X_{n-1}, \ldots, X_{n-r}) \pmod{N}$$

must be ultimately periodic with period not more than N^{r+1}.

10. In the model of Wyner's wiretap channel, suppose that both the main channel and the wiretap channel are binary symmetric channels. Prove that

$$\Gamma(R) = C_M - C_{MW} \quad (0 \le R \le C_M),$$

where C_M is the capacity of the main channel and C_{MW} is the capacity of the cascaded combination of the main and wiretap channels.

(Leung Yan Cheong, 1977)

11. If the main channel is binary symmetric with bit error probability p, and the wiretap channel is also binary symmetric with bit error probability $p' > p$, what is the secret capacity of the system?

Appendix 1

Proof of the uniqueness theorem that $H = -\sum p_i \log p_i$

Let H satisfy (A1)–(A8) of Chapter 1, and define

(1) $$g(n) = H(1/n, 1/n, \ldots, 1/n) \quad (n \in \mathbb{Z}).$$

From (A7),

$$g(n^k) = g(n) + g(n^{k-1})$$

and hence

(2) $$g(n^k) = kg(n) \quad (k, n \in \mathbb{Z}).$$

Now, for $r, s \in \mathbb{R}$ and $n \in \mathbb{Z}^+$, let m satisfy

(3) $$r^m \le s^n \le r^{m+1};$$

then, using (2) and the monotonicity of g (from (A5)), we have

$$g(r^m) \le g(s^n) \le g(r^{m+1}),$$

so

$$mg(r) \le ng(s) \le (m+1)g(r).$$

Combining this with the rewriting of (3) in the form

$$m \ln(r) < n \ln(s) < (m+1)\ln(r)$$

gives

$$\left| \frac{g(s)}{g(r)} - \frac{\ln(s)}{\ln(r)} \right| \le \frac{1}{n}.$$

Since n was an arbitrary member of $\mathbb{Z}$, the right-hand side can be made arbitrarily small, showing

(4) $$g(s)/\ln(s) = g(r)/\ln(r) = A,$$

where A is some constant. This shows that

(5) $$g(s) = A \ln(s) \quad (s \in Z).$$

Next suppose p $(0 < p < 1)$ is a rational, say $p = t/n$, with t and n integers and $q = (n - t)/n$. Condition (A8) gives

$$g(n) = H\left(\frac{1}{n}, \ldots, \frac{1}{n}\right) = H\left(\frac{t}{n}, \frac{n-t}{n}\right) + \frac{t}{n}g(t) + \frac{n-t}{n}g(n-t).$$

Using (5) and collecting terms gives

$$H\left(\frac{t}{n}, \frac{n-t}{n}\right) = A\left(\frac{t}{n}\right)\ln\frac{t}{n} + A\left(\frac{n-t}{n}\right)\ln\frac{n-t}{n}.$$

This gives

(6)
$$H(p, 1-p) = Ap \ln p + A(1-p)\ln(1-p),$$

whenever p is rational. The continuity assumption (A6) extends (6) to all p $(0 < p < 1)$.

To complete the proof that, for any $p_i > 0$ with $p_1 + \ldots + p_N = 1$, we have

(7)
$$H(p_1, \ldots, p_N) = A \sum_1^N p_i \ln p_i,$$

we use induction on N. We have already shown (7) is true for $N = 2$. Assume (7) for N and consider $H(p_1, \ldots, p_{N+1})$. We let

$$p = p_1 + \ldots + p_N, \qquad q = p_{N+1},$$

and apply (A8).
This gives

$$H(p_1, \ldots, p_{N+1}) = H(p, q) + pH\left(\frac{p_1}{p}, \ldots, \frac{p_N}{p}\right) + qH(1)$$

$$= Ap \ln p + Aq \ln q + pA \sum_{i=1}^N \frac{p_i}{p} \ln \frac{p_i}{p},$$

using the inductive hypothesis. Writing the last equality as

$$H(p_1, \ldots, p_{N+1}) = Ap \ln p + Ap_{N+1} \ln p_{N+1} + A \sum_{i=1}^N p_i(\ln p_i - \ln p),$$

and remembering that $\sum_1^N p_i = p$, we obtain

$$H(p_1, \ldots, p_{N+1}) = A \sum_{i=1}^{N+1} p_i \ln p_i,$$

as required. $\qquad\qquad\qquad\qquad\qquad\qquad\qquad\qquad\qquad\qquad\square$

Appendix 2

Letter frequencies of English

Frequency of letters in 1000 characters of English

A	64		N	56
B	14		O	56
C	27		P	17
D	35		Q	4
E	100		R	49
F	20		S	56
G	14		T	71
H	42		U	31
I	63		V	10
J	3		W	18
K	6		X	3
L	35		Y	18
M	20		Z	2

SPACE/PUNCTUATION MARK 166

Figures taken from the copy-fitting tables used by professional printers based on counts involving hundreds of thousands of words in a very wide range of English language printed matter.

Answers to exercises

Chapter 1

   1.1 Both have entropy $\frac{1}{2} \log 48$.

   3.1 Any random variable taking values $\pm a$ with equal probability.

   4.1 The sequence of 26 digits whose information content is $26 \log 10$.

Chapter 2

   2.1 With $m = 4$ say, code is 1, 01, 001, 0001.

   3.1 $2^7 = 128$.

   4.1 n must be a power of 2.

   4.2 Average length is $k - (2^k - 1)^{-1}$ and $H = k + \log(1 - 2^{-k})$.

   5.1 2.4 bits.

   5.2 29/12 bits for code

$$00, 01, 11, 101, 1000, 1001$$

Lower bound 2.36 bits. Upper bound 3.36 bits.

   5.3 Code is

$$1, 2, 00, 02, 010, 011.$$

Chapter 3

   2.1 There must be either 2 errors or 4 errors in transmission of word.

   3.1 Ideal observer decodes as c_1.

         Maximum likelihood decodes as either c_1 or c_3 arbitrarily.

   4.1 Capacity is $1 - \varepsilon$ bits.

   5.1 570 bits.

   5.2 Capacity $> 5/8$ bits $\Rightarrow p < 0.0725$ or $p > 0.9275$.

Chapter 4

3.1 $\begin{pmatrix} 1 & 0 & 0 & 1 & 1 \\ 0 & 1 & 0 & 1 & 1 \\ 0 & 0 & 1 & 0 & 0 \end{pmatrix}$

3.2 Code has 8 codewords and minimum distance 1.

3.3 $[n_1 + n_2, k, d_3]$ where $d_3 \geq d_1 + d_2$.

4.1 0000000 1101010

4.2 1000111 0011110

4.3 $n = 7$, $M = 16$, $d = 3$.

5.1 (a) 1110011   (b) 1101011
    (c) 0110110   (d) 1111000.

6.1 H is the 4 by 15 matrix with its ith column the binary representation of i $(1 \leq i \leq 15)$.
    (a) 000 000 000 000 000
    (b) 111 111 111 111 110.

7.1 $\mathscr{C}_1$ has generator polynomial $1 + x + x^2$ and has 2 codewords
    $\mathscr{C}_2$ has generator polynomial $1 + x$ and has 4 codewords
    $\mathscr{C}_3$ has generator polynomial 1 and has 8 codewords
    $\mathscr{C}_4$ is the trivial code with just the zero codeword.

7.2 $1 + x + x^3$.

Chapter 5

1.1 $-(p \log p + q \log q)$.

1.2 Yes: it has entropy $H = \frac{1}{2}$.

2.2 Only the source of Exercise 2.

3.1 $2^{(\frac{2}{3} + \log 5 - \frac{1}{10} \log 3)n} = (4.75)^n$

3.2 $\log_2 10 \approx 3.3$ bits.

5.1 $H = \frac{10}{13} + \frac{3}{13} \log 3$.

5.2 (a) and (d).

5.3 $a_1 = \frac{3}{13}$, $a_2 = \frac{4}{13}$, $a_3 = \frac{6}{13}$.

Chapter 6

1.2 5.56 letters.

2.1 $(2.83)^N$

3.1 Finite vocabulary unless $0 \leq D < 1$.

4.1 MATHEMATICS IS A PRETTY EXACT SCIENCE.

4.2 FACT IS STRANGER THAN FICTION.

Chapter 7

1.1 Decryption does not give a unique message.

1.2 25!

2.1 d^2

2.2 (a) Any message of $|\Sigma| - 1$ distinct symbols from Σ.
(b) Any message of length d.

4.2 $d_1!d_2!$

4.3 $pH_1 + (1-p)H_2$.

5.2 Transposition ciphers preserve frequency of letters so $H(C_N)$ is more accurately given by the first-order entropy rather than zero-order entropy.

Chapter 8

2.1 Period is 4.

4.2 It is a $(3, 3, 2)$ code.

Chapter 9

1.1 Show that in the worst case any algorithm must examine *every* entry of the $n \times n$ matrix.

1.2 Brute force count is $O(mn)$, its difficult to envisage a faster algorithm.

1.3 (Hint) Form the binary expansion of n.

2.1 Standard methods of Gaussian elimination are $O(n^3)$ so problem is in P.

2.2 Brute force examination of all possible k-sets of vertices is $O(n^k)$.
[It is possible by a clever method based on fast matrix multiplication to bring this exponent down to $O(n^{\alpha k})$, $\alpha > 0$ but whether it can be brought down further say to $O(n^3)$ seems to be a difficult problem.]

3.2 (a) determinant $\in P$. Hence the problem is in NP.
(b) permanent $> n \in$ NP.
(c) as far as I know no one can show that this problem $\in$ NP.

3.3 (a), (b) are in NP.
(c) is not known to be in NP.

4.1 Use the transformation $G \rightarrow \bar{G}$ where u, v are joined by edge in $\bar{G}$ iff not in G.

4.2 Yes: LONGPATH $\in$ NP and HAMILTON $\propto$ LONGPATH.

5.2 List of Boolean functions mapping x, y into $f(x, y)$.

	$f(x, y)$	Symbol
PROJECTION	x	π_1
PROJECTION	y	π_2
CONJUNCTION (AND)	xy	$\wedge$
DISJUNCTION (OR)	$x + y + xy$	$\vee$
IMPLICATION	$1 + x + xy$	$\rightarrow$
IMPLICATION	$1 + y + xy$	$\leftarrow$
NOT-EQUIVALENCE	$x + y$	$\oplus$
CONSTANT	0	0

and their negations, obtained by adding the Boolean variable 1 to each of the above. For example

$$\bar{\pi}_1 = \bar{x} = 1 + x.$$

Chapter 10

1.1 $518940557 = 15107 \times 34351.$
Factoring a 200 digit number on a machine with 1 µs instruction time could take at least a million years.

1.2 Take $A = \mathrm{diag}(p, p)$, $B = \mathrm{diag}(q, q)$ where p, q are large primes.

2.1 $C = 110111$

2.2 $M = 101110$

4.3 Take $n = 5$, $a = 4$, $y = 4$, then $x = 1$ or 3.

5.1 6.

5.2 3.

6.1 $M = 2.$

Chapter 11

2.1 $e = 5$, $M = 5.$

2.2 3031.

3.2 CURE.

4.1 M is one of 48, 15, 100, 67.

5.2 $C_2 = 29.$

Chapter 12

2.1 Send $23^{17} \bmod 77 = 67.$

2.2 Send one of 12, 100, 107, 195.

3.1 (a) $(X_1, Y_1, X_1^2 \bmod 13, \ Y_1^7 \bmod 13)$ and $(X_2, Y_2, X_2^5 \bmod 13,$ $Y_2^3 \bmod 13)$ where $X_i, \ Y_i$ are chosen at random from $\{1, 2, \ldots, 13\}$.

(b) With the signature 75.

3.2 $M_1 = 1, \ M_2 = 3.$

4.1 $S = 4^3 \bmod 15 = 4.$

5.1 S_1 is one of 3, 17, 18, 32.
S_2 is one of 8, 13, 22, 27.

5.2 With either $2S$ or $(N-2)S$ where N is the modulus of the scheme.

Chapter 13

1.1 $2^{26} - 2.$

2.2 0, 1.

4.2 $-p \log p - q \log q.$

Hints and answers to problems

Chapter 1

1. I (1000 words) $\simeq 13\,288$ bits.
   I (TV picture) $\simeq 1\,200\,000$ bits.
2. Not true: take $g(X, Y)$ to be a constant and $H(g(X, Y) \mid X) = 0$.
4. 0.012 bits.
5. Express conditional entropies as joint entropies by using Theorem 3.2.
6. Write $(X_1, \ldots, X_{n+1})$ as (U, X_{n+1}) and use Theorem 3.2.
8. A binomial random variable is the sum of independent random variables taking the values 0 and 1 with probabilities p and q.
11. Use method of Lagrange multipliers.

Chapter 2

1. Use noiseless coding theorem; the existence of a 5 question winning strategy would contradict it.
2. $\lceil 2 \log n \rceil$
3. 2.45 bits
4. Maximum amount of information from one weighing is $\log 3$ bits.
5. If $\mathscr{C}$ is a uniquely decipherable code then it can be used when the source words are equiprobable.
6. Same idea as questions 1 and 2: identify a strategy with a binary decision tree.
7. This is hard. Try first the case $D = 2$ to get the idea. For general D write $N = D + k(D - 1)$ and use induction on k.

Chapter 3

2. $(N - k + 1)p^k$
3. $(35\varepsilon - 54\varepsilon^2 + 13\varepsilon^3 + 6\varepsilon^4)/60$

4. The probability of error in the cascade is p_N.
5. Look at the proof of capacity of the extended channel in Section 3.1.
7. The binary symmetric channel with error probability $\frac{1}{2}$, that is the perfect scrambler, is one example.
8. The entropy of the output conditional on the input is independent of the input.
9. $2^{-n} \sum_{r=k}^{n} \binom{n}{r}$
12. Use the tail inequality.
13. 33.3 per second.

Chapter 4

1. Identify alphabet of size N with integers $0, 1, \ldots, N-1$ and use 'parity' digit to make sum congruent to $0 \bmod N$.
2. More than 2^{300} years even using (4.1.2).
4. If $\mathscr{C}$ is optimum (n, d)-code divide its words into those ending in 0 or 1. One of these sets gives $A_2(n-1, d) \geq A_2(n, d)/2$.
5. Alter first bit of codewords of odd weight in the $(n, M, 2k)$-code.
6. $q^7 + 5q^6p + 2q^5p^2$ where $q = 1-p$.
7. Minimum distance equals minimum weight of non-zero codeword.
8. (ii) Remove last $(d-1)$ columns from the matrix whose rows are the codewords.
9. Condition is $\sum_{i=0}^{3} \binom{n}{i} = 2^t$ for some $t < n$, which reduces to

$$(n+1)[(n+1)^2 - 3(n+1) + 8] = 3.2^{t+1}.$$

Now show $n+1$ not divisible by 16.
13. $1 - \sum_{i=0}^{t} \binom{n}{i} p^i (1-p)^{n-i}$.
15. A generator g of $\mathscr{C}$ must divide $(x-1)(x^{n-1} + x^{n-2} + \cdots + 1)$ and if g is a multiple of $(x-1)$ then every codeword has even weight.
17. Start from the Hamming $[15, 11, 3]$-code and use Exercise 4.3.3. It has 2^{11} codewords and will correct up to 2 errors.
18. If $g_0 = 0$ then $g_1 + g_2 x + \cdots + g_k x^{k-1}$ is a generator polynomial and is of lower degree.
22. Consider the M by n matrix A whose rows are the codewords. The sum of the distances is

$$\sum_{i=1}^{n} 2x_i(M - x_i)$$

where x_i is the number of zeros in the ith column. Maximising this sum over the x_i gives the right hand inequality. The left hand inequality is trivial.

24. Probability of error is 0.14×10^{-4}; Rate $= \frac{3}{16}$.

Chapter 5

1. Either 0.14 or 0.86.
2. $H(\mathcal{S}) = \frac{3}{2}$; $\mathcal{S}_1$ is not stationary; $4H(\mathcal{S}_1) = H(\mathcal{S})$.
3. $(88 + 27 \log 3)/95$.
4. $(a_1 + a_2) + (a_3 + a_4)(8 - 3 \log 3)/4$ where (a_1, a_2, a_3, a_4) is the initial distribution.
7. $(44 + 15 \log 3 + 10 \log 5)/360$.

Chapter 6

2. $\frac{2}{9}$.
3. If p_i is the probability of i then $p_i = \sum_j p_{ij}$ where p_{ij} is the digram probability.
6. Use the approximation $\sum_{r=1}^{N} 1/r \simeq \ln(N)$.
7. Only if you assume source emitting language is memoryless.
8. Use the same approximation as in Problem 6.

Chapter 7

1. Both equal log 10; unicity is infinite.
3. (a) $d - 1$ (b) d^2.
4. Almost any example in which the set of cryptograms of the first system forms a very small subset of the message space of the second cryptosystem will contradict the second inequality.
6. The empirical value for U seems to be about 3.5 whereas using the unicity formula gives 1.46. The discrepancy is because the Caesar is so weak a system we should take a low order entropy for H not H_E. The two pieces of text proposed by S. H. Babbage are

WEARINGAM AUVEHATHEFISHESAU STERELY
THEMANWITHTHECA KEFORKDROPSCROCKE RYREGULARLY

A more doubtful piece of prose with an 'ambiguity length' of 31 symbols has also been constructed by Babbage.

9. $(3.7)^{-1}\log(d!) \le U \le (2.7)^{-1}\log(d!)$.
10. $d(d+1)$.
11. (a) $2^d d!$ keys.
12. 300.
13. $1.47\,d$ symbols.

Chapter 8

2. A four register machine with $c_1 = c_2 = 0$, $c_3 = c_4 = 1$, initial vector $x(0) = (0, 1, 1, 0)$.
3. Use the finiteness of the system.
5. Yes: the difference equation (only) determines the coefficients of the shift register.
6. $c_1 = c_2 = c_3 = 0$, $c_4 = c_5 = 1$.
7. In general n terms are needed.
8. Yes: consider the 5 state machines having

$$c_1 = c_2 = c_3 = 0, \ c_4 = c_5 = 1$$
$$c_1' = 1 = c_5', \ c_2' = c_3' = c_4' = 0$$

with initial state 01101 they produce the same sequence.
9. No: $u \sim v$ does not imply $v \sim u$.
10. $x(0) = (C')^{-1}a$.
11. Consider the number of ways of constructing a non-singular matrix by successively picking row vectors which are not linearly dependent on those already picked.
13. $1 + x^5 + x^6$ is one example.
14. $G(x)(1 - x^p)$ is a polynomial in x of finite degree.

Chapter 9

1. $O(n^2)$.
2. $O(2^{nA})$ where A is between 72 and 79 (see 4.1.2).
4. If $c(n)$ is number of comparisons show

$$c(2^k) = 2c(2^{k-1}) + 2$$

giving

$$c(2^k) = 3.2^{k-1} - 2.$$

5. Relate to the complementary graph $\bar{G}$ which has the same set of

vertices as G but an edge joins x, y in $\bar{G}$ only if x, y are unjoined in G.

6. Use Turing machine representation of NP; all possible certificates can be checked in time $2^{p(n)}$.

7. The obvious circuit has 5 gates but there is a circuit with only one gate.

12. Complexity is about $100 \lceil \log n \rceil^2$.

Chapter 10

1. Only the composition (c).

5. If not, the algorithm which finds the 'easy' bits first and the 'hard' bits by exhaustive search is a polynomial time algorithm.

6. Problem reduces to solving linear equations in the entries of A.

7. Yes: exhaustive search and checking all possibilities.

8. Represent the transformation as multiplication by a permutation matrix.

10. Take a fixed message M and consider the function f on the key space which maps a key K to the encryption of M under K.

11. Use the Euler–Fermat identity.

12. $D = K^{-1}$ may not exist, e.g. $p = 7$, $K = 2$.

Chapter 11

1. 47.

4. $O(2^n)$.

5. Simple counting argument.

7. Use the Euler–Fermat identity.

9. (i) Show $x^2 = 1 \pmod{n}$ has two solutions.
   (ii) $d = 6$, $n = 15$.

Chapter 12

1. With the integer 225.

2. With $(12, 8, 8, 1)$.

3. The signature is not unique and therefore not checkable.

4. $d = 15551$.

Chapter 13

1. $N!/(m!)^t$.
2. x ends in 0 only if C is a quadratic residue mod p.
3. Show that a is a non-residue if and only if $a^{p-1} = -1 \pmod p$.
4. See Theorem 9.6.2.
5. 104 625.
6. Pick y at random from Z_N and test whether y is a non-residue. Average number of choices before success is 4.
9. Counting argument.
11. $C_S = p \log p + q \log q - r \log r - s \log s$ where $r = pp' + qq'$ and $s = 1 - r$.

References

Abramson, N. M. (1963). *Information theory and coding*. McGraw-Hill, New York.

Aczél, J. and Daróczy, Z. (1975). *On measures of information and their characterizations*. Academic Press, New York.

Adleman, L. M. (1978). Two theorems on random polynomial time. *IEEE FOCS* **19,** 75–83.

Adleman, L. M. (1979). A subexponential algorithm for the discrete logarithm problem with applications to cryptography. *IEEE FOCS* **20,** 55–60.

Adleman, L. M. (1983). On breaking the iterated Merkle–Hellman public key cryptosystem. In Chaum, D. *et al.* (eds.) *Advances in Cryptology: Proceedings of Crypto* **82,** 303–8.

Adleman, L. M. and Manders, K. (1977). Reducibility, randomness and intractability. *Proc. 9th ACM Symp. Theory Comp.* 151–3.

Adleman, L. M., Pomerance, C., and Rumely, R. S. (1983). On distinguishing prime numbers from composite numbers. *Ann. Math.* **117,** 173–206.

Aho, J. D., Hopcroft, J. E., and Ullman, J. D. (1974). *The design and analysis of computer algorithms*. Addison-Wesley, Reading, Mass.

Alexi, W., Chor, B., Goldreich, O., and Schnorr, C. P. (1984). RSA/RABIN bits are $1/2 + 1/poly(\log N)$ secure. *IEEE FOCS* **25,** 449–57.

Ash, R. (1965). *Information theory*. Wiley, New York.

Asmuth, C. A. and Blakley, G. R. (1981). An efficient algorithm for constructing a cryptosystem which is harder to break than two other cryptosystems. *Comp. and Maths. with Appls.* **7**(6), 447–50.

Babbage, S. H. (1987). (Private communication.)

Bach, E. (1985). *Analytic methods in the analysis and design of number-theoretic algorithms*. MIT Press, Cambridge, Mass.

Beker, H. and Piper, F. (1982). *Cipher systems*. Van Nostrand Reinhold, London.

Ben-Or, M., Chor, B., and Shamir, A. (1983). On the cryptographic security of single RSA bits. *Proc. 15th Symp. Theory Comp.* 421–30.

Berlekamp, E. R. (1968). *Algebraic coding theory*. McGraw-Hill, New York.

Berlekamp, E. R., McEliece, R. J., and van Tilburg, H. C. A. (1978). On the inherent intractability of certain coding problems. *IEEE Trans. Info. Theory* **24,** 384–6.

Billingsley, P. (1965). *Ergodic theory and information.* Wiley, New York.

Billingsley, P. (1979). *Probability and measure.* Wiley, New York.

Blahut, R. E. (1983). *Theory and practice of error control codes.* Addison-Wesley, Reading, Mass.

Blake, I. F. and Mullin, R. C. (1976). *An introduction to algebraic and combinatorial coding theory.* Academic Press, New York.

Blake, T., Fuhi-Hara, R., Mullin, R., and Vanstone, S. (1984). Computing logarithms in finite fields of characteristic two. *SIAM J. Alg. Discr. Methods* **5,** 276–85.

Blum, M. and Goldwasser, S. (1985). An efficient probabilistic public key encryption scheme which hides all partial information. *Advances in cryptology: Proc. Crypto* **84,** 289–99.

Blum, M. and Micali, S. (1982). How to generate cryptographically strong sequences of pseudo-random bits. *IEEE FOCS* **23,** 112–17. (*Extended version appeared in SIAM J. Comp.* **13**(4), 850–64.)

Blum, L., Blum, M., and Shub, M. (1983). Comparison of two pseudo-random number generators. In Chaum, D. *et al.* (eds.) *Advances in cryptology: Proc. Crypto* **82,** 61–79.

Brassard, G. (1979). A note on the complexity of cryptography. *IEEE Trans. Info. Theory* **IT-25**(2), 232–4.

Brickell, E. F. and Moore, J. M. (1983). Some remarks on the Herlestam–Johannesson algorithm for computing logarithms over $GF(2^P)$. In Chaum, D. *et al.* (eds.) *Advances in cryptology: Proc. Crypto* **82,** 15–19.

Bright, H. S. and Enison, R. L. (1979). Quasi-random number sequences from a long-period TLP generator with remarks on application to cryptography. *ACM Computing Surveys* **11**(4), 357–70.

Burton, N. G. and Licklider, J. C. R. (1955). Long-range constraints in the statistical structure of printed English. *Amer. J. Psych.* **68,** 650–3.

Cameron, P. J. and van Lint, J. h. (1980). *Graphs, codes and designs.* London Math. Soc. Lecture Note Series, Vol. 43, Cambridge University Press, Cambridge.

Carleial, A. B. and Hellman, M. E. (1977). A note on Wyner's wiretap channel. *IEEE Trans. Info. Theory* **IT-23**(3), 387–90.

Chapanis, A. (1954). The reconstruction of abbreviated printed messages. *J. Experimental Psych.* **48**(6), 496–510.

Cohen, H. and Lenstra, H. W. (1984). Primality testing and Jacobi sums. *Math. Comp.* **42,** 297–330.

Cook, S. A. (1971). The complexity of theorem-proving procedures. *Proc. 3rd ACM Symp. Theory Comp.* 151–8.

Coppersmith, D. (1984). Fast evaluation of logarithms in fields of characteristic two. *IEEE Trans. Info. Theory* **IT-30,** 587–94.

Coppersmith, D. and Winograd, S. (1987). Matrix multiplication via arithmetic progressions. *J. Symb. Algebra* (to appear).

Cover, T. M. and King, R. C. (1977). A convergent gambling estimate of the entropy of English. *IEEE Trans. Info. Theory* **IT-24,** 413–21.

Csiszár, I. and Körne, J. (1981). *Information theory: coding theorems for discrete memoryless systems.* Academic Press, New York.

Davis, J. A., Holdridge, D. B., and Simmons, G. J. (1985). Status report on factoring (at the Sandia National Laboratories). *Advances in cryptology: Proc. Crypto* **84,** 183–215.

Davis, R. M. (1978). The Data Encryption Standard in perspective. *IEEE Comm. Soc. Magazine* **16,** 5–9.

Deavours, C. A. (1977). Unicity points in cryptanalysis. *Cryptologia* **1**(1), 46–68.

DeLaurentis, J. M. (1984). A further weakness in the common modulus protocol for the RSA cryptoalgorithm. *Cryptologia* **8**(3).

Denning, D. (1983). *Cryptography and data security.* Addison-Wesley, Reading, Mass.

Diffie, W. and Hellman, M. E. (1976). A critique of the proposed Data Encryption Standard. *Comm. ACM* **19**(3), 164–5.

Diffie, W. and Hellman, M. E. (1976). New directions in cryptography. *IEEE Trans. Info. Theory* **22,** 644–54.

Diffie, W. and Hellman, M. E. (1977). Exhaustive cryptanalysis of the NBS Data Encryption Standard. *Computer* **10**(6), 74–84.

Diffie, W. and Hellman, M. E. (1979). Privacy and authentication: an introduction to cryptography. *Proc. IEEE* **67**(3), 397–427.

Dixon, J. D. (1981). Asymptotically fast factorization of integers. *Math. Comp.* **36,** 255–60.

Dixon, J. D. (1984). Factorisation and primality tests. *Amer. Math. Monthly,* June, 333–52.

Elgamal, T. (1985). A public key cryptosystem and a signature scheme based on discrete logarithms. *IEEE Trans. Info. Theory* **31,** 469–72.

Elias, P. (1955). Coding for noisy channels. *IRE Convention Record* **4,** 47–46.

Evans, A., Kantrowitz, W., and Weiss, E. (1974). A user authentication scheme not requiring secrecy in the computer. *Comm. ACM* **17**(8), 437–42.

Fano, R. M. (1961). *Transmission of information.* MIT Press, Cambridge, Mass.

Feinstein, A. (1958). *Foundations of information theory.* McGraw-Hill, New York.

Feinstein, A. (1959). On the coding theorem and its converse for finite-memory channels. *Information and Control* **2**(1), 25–44.

Feistel, H. (1973). Cryptography and computer privacy. *Scientific American* May, 15–23.

Feistel, H., Notz, W. A., and Smith, J. L. (1975). Some cryptographic techniques for machine-to-machine data communications. *Proc. IEEE* **63**(11), 1545–54.

Fischer, E. (1979). Language redundancy and cryptanalysis. *Cryptologia* **3,** 233–5.

Friedman, W. F. (1973). Cryptology. *Encyclopaedia Britannica,* p. 848.

Frieze, A. M., Kannan, R., and Lagarias, J. C. (1984). Linear congruential generators do not produce random sequences. *IEEE Focs* **25**, 480–4.

Gaines, H. F. (1956). *Cryptanalysis, a study of ciphers and their solution.* Dover, New York.

Gallager, R. G. (1968). *Information theory and reliable communication.* Wiley, New York.

Gallager, R. G. (1978). Variations on a theme by Huffman. *IEEE Trans. Info. Theory* **IT-24**, 668–74.

Gardner, M. (1977). Mathematical games. *Scientific American,* August, 120–4.

Garey, M. and Johnson, D. (1979). *Computers and intractability: a guide to the theory of NP-completeness.* Freeman, San Francisco.

Geffe, P. R. (1967). An open letter to communication engineers. *Proc. IEEE* **55**, 2173.

Gilbert, E. N. (1952). A comparison of signalling alphabets. *Bell Syst. Tech. J.* **31**, 504–22.

Givierge, M. (1925). *Cours de cryptographie.* Paris (2nd edn., 1932).

Golay, M. J. E. (1949). Notes on digital coding. *Proc. IEEE* **37**, 657.

Golay, M. J. E. (1954). Binary coding. *IEEE Trans. Info. Theory* **4**, 23–8.

Goldreich, O., Goldwasser, S., and Micali, S. (1984). How to construct random functions. *IEEE Focs* **25**, 464–479.

Goldwasser, S. and Micali, S. (1982). Probabilistic encryption. *IEEE FOCS* **23**. (Later version appeared in *J. Comp. Syst. Sci.* **28**(2), 270–99.

Goldwasser, S. and Kilian, J. (1986). Almost all primes can be quickly certified. *Proc. 18th ACM Symp. Theory Comp.* 316–30.

Goldwasser, S., Micali, S., and Tong, P. (1982). Why and how to establish a private code on a public network. *IEEE FOCS* **23**, 134–44.

Goldwasser, S., Micali, S., and Yao, A. (1983). Strong signature schemes. *Proc. 15th ACM Symp. Theory Comp.* 431–39.

Goldwasser, S., Micali, S., and Rivest, R. L. (1984). A 'paradoxical' solution to the signature problem. *IEEE* 441–48.

Golomb, S. W. (1967). *Shift register sequences.* Holden-Day, San Francisco.

Good, I. J. (1969). See Meetham, A. R. (ed.) *Encyclopaedia of Linguistics,* Pergamon Press, Oxford.

Goppa, V. D. (1970). A new class of linear error-correcting codes. *Problems of Info. Transmission* **6**(3), 207–12.

Gray, R. and Davisson, L. D. (1977). *Ergodic and information theory: benchmark papers in electrical engineering and computer science.* Dowden, Hutchinson and Ross.

Grimmett, G. R. and Stirzaker, D. R. (1982). *Probability and random processes.* Oxford University Press, Oxford.

Grimmett, G. R. and Welsh, D. J. A. (1986). *Probability: an introduction.* Oxford University Press, Oxford.

Grollmann, J. and Selman, A. (1984). Complexity measures for public key cryptosystems. *IEEE FOCS* **25**, 495–503.

Guiasu, S. (1977). *Information theory with applications.* McGraw-Hill, New York.

Guiasu, S. and Shenitzer, A. (1985). The principle of maximum entropy. *Math. Intelligencer* **7**, 42–8.

Hamming, R. W. (1950). Error detecting and error correcting codes. *Bell Syst. Tech. J.* **29**, 147–60.

Hamming, R. W. (1980). *Coding and information theory.* Prentice-Hall, New Jersey.

Hartley, R. V. L. (1928). Transmission of information. *Bell Syst. Tech. J.* **7**, 535–63.

Hellman, M. E. (1977). An extension of the Shannon theory approach to cryptography. *IEEE Trans. Info. Theory* **IT–23**(3), 289–94.

Hellman, M. E. (1978). An overview of public-key cryptography. *IEEE Trans. Commun.* **16**(6), 24–32.

Hellman, M. E. (1979). DES will be totally insecure within ten years. *IEEE Spectrum* **16**, 32–9.

Hellman, M. E. (1980). A cryptanalytic time-memory trade off. *Trans. IEEE Info. Theory* **IT–26**(4), 401–6.

Hellman, M. E. and Merkle, R. C. (1981). On the security of multiple encryption. *Comm. ACM* **24**, 465–7.

Hellman, M. E. and Reyneri, J. M. (1983). Fast computation of discrete logarithms in *GF*(2). In Chaum, D. *et al.* (eds.) *Advances in cryptology: Proc. Crypto* **82**, 3–13.

Hill, L. S. (1929). Cryptography in an algebraic alphabet. *Amer. Math. Monthly* **36**, 306–12.

Hill, L. S. (1931). Concerning certain linear transformation apparatus of cryptography. *Amer. Math. Monthly* **38**, 135–54.

Hill, R. (1986). *A first course in coding theory.* Oxford University Press, Oxford.

Hille, E. and Phillips, R. S. (1957). *Functional analysis and semi-groups.* Amer. Math. Soc. Colloq. Publ. **31**.

Hopcroft, J. E. and Ullman, J. D. (1979). *Introduction to automata theory, languages, and computation.* Addison-Wesley, Reading, Mass.

Huffman, D. A. (1952). A method for the construction of minimum redundancy codes. *Proc. IRE* **40**(10), 1098–1101.

Jurgensen, H. (1983). Language redundancy and the unicity point. *Cryptologia* **7**, 37–48.

Kahn, D. (1967). *The codebreakers: the story of secret writing.* Macmillan, New York.

Karp, R. M. (1972). Reducibility among combinatorial problems. In Miller, R. E. and Thatcher, J. W. (eds.) *Complexity of computer computations.* Plenum Press, New York, pp. 85–103.

Karush, J. (1961). A simple proof of an inequality of McMillan. *IRE Trans. Info. Theory* **IT–7**(2), 118.

Keyes, R. W. (1975). Physical limits in digital electronics. *Proc. IEEE* **63**, 740–67.

Khinchin, A. (1957). *Mathematical foundations of information theory.* Dover Publications, New York.

Knuth, D. E. (1969). *The art of computer programming: seminumerical algorithms,* vol. 2. Addison-Wesley, Reading, Mass.

Komamiya, Y. (1953). Application of logical mathematics to information theory. *Proc. 3rd Japan. Nat. Cong. Appl. Math.,* 437.

Konheim, A. G. (1981). *Cryptography: a primer.* Wiley New York.

Kranakis, E. (1986). *Primality and cryptography.* Wiley, New York.

Kullback, S. (1976). *Statistical methods in cryptanalysis.* Aegean Park Press.

Ladner, R. E. (1975). On the structure of polynomial time reducibility. *J. Assoc. Comp. Mach.* **22,** 155–71.

Lamport, L. (1979). Constructing digital signatures from a one-way function. *SRI Intl. CSL* **98.**

Landauer, R. W. (1961). Irreversibility and heat generation in the computing process. *IBM J. Res. Develop.* **5,** 183–91.

Lempel, A. (1979). Cryptology in transition. *ACM Computing Surveys* **11**(4), 285–303.

Lenstra, H. W. (1980–81). Primality testing algorithms. *Séminaire Bourbaki* **33,** 243–57.

Leung-Yan-Cheong, S. K. (1977). On a special class of wiretap channels. *IEEE Trans. Info. Theory* **IT–23,** 625–7.

Levenshtein, V. I. (1961). The application of Hadamard matrices to a problem in coding. *Probl. Kibernet.* **5,** 123–136. English translation in *Problems of Cybernetics* **5** (1964), 166–84.

Lidl, R. and Niederreiter, H. (1986). *Introduction to finite fields and their applications.* Cambridge University Press, Cambridge.

van Lint, J. H. (1982). *Introduction to coding theory.* Springer-Verlag, New York.

Luby, M. and Rackoff, C. (1986). Pseudo-random permutation generators and cryptographic composition. *Proc. 18th Symp. Theory Comp.* 356–63.

McEliece, R. J. (1977). *The theory of information and coding.* Addison-Wesley, Reading, Mass.

McEliece, R. J. (1978). A public-key cryptosystem based on algebraic coding theory. *DSN Progress Report,* 42–4.

McMillan, B. (1953). The basic theorems of information theory. *Ann. Math. Stat.* **24**(2), 196–219.

MacWilliams, F. J. and Sloane, N. J. A. (1977). *The theory of error-correcting codes.* North-Holland, Amsterdam.

Mandelbrot, B. (1953). An informational theory of the statistical structure of language. In Jackson, W., *Communication theory,* Academic Press, New York, pp. 486–502.

Mandelbrot, B. (1954). Structure formelle des textes et communication. *Word* **10,** 1–26.

Mandelbrot, B. M. (1977). *Fractals, form chance and dimension.* Freeman, San Francisco.

Matyas, S. (1979). Digital signatures—an overview. *Computer Networks* **3**, 87–94.

Merkle, R. C. (1978). Secure communications over insecure channels. *Comm. ACM* **21**(4), 294–9.

Merkel, R. C. (1979). *Secrecy, authentication and public key systems*. Ph.D. dissertation, Dept. of Electrical Engineering, Stanford University.

Merkel, R. C. (1979). *Secrecy, authentication and public key systems*. Bowker Publ., Epping.

Merkle, R. C. and Hellman, M. E. (1978). Hiding information and signatures in trapdoor knapsacks. *Trans. IEEE Info. Theory* **IT–24**(5), 525–530.

Meyer, C. and Matyas, S. (1982). *Cryptography*. Wiley, New York.

Miller, G. A. (1981). *Language and speech*. Freeman, San Francisco.

Miller, G. A. and Friedman, E. A. (1957). The reconstruction of mutilated English texts. *Info. Contr.* **1**, 38–55.

Miller, G. A., Newman, E. B., and Friedman, E. A. (1957). Length-frequency statistics for written English. *Info. Contr.* **1**, 370–89.

Miller, J. (1975). On factorization with a suggested new approach. *Math. Comp.* **29**, 155–72.

de Millo, R. A., Dobkin, D. P., Jones, A. K., and Lipton, R. J. (1978). *Foundations of secure computation*. Academic Press.

Morris, R. (1978). The Data Encryption Standard—retrospective and prospects. *IEEE Comm. Soc. Magazine* **16**(6), 11–14.

Morris, R., Sloane, N. J. A., and Myner, A. D. (1977). Assessment of the National Bureau of Standards' proposed federal Data Encryption Standard. *Cryptologia* **1**, 281–306.

Newman, E. and Waugh, N. (1960). The redundancy of texts in three languages. *Info. Contr.* **3**, 141–53.

Odlyzko, A. M. (1985). Discrete logarithms in finite fields and their cryptographic significance. *Proc. Crypto* **84**, 225–314.

Patterson, N. J. (1975). The algebraic decoding of Goppa codes. *IEEE Trans. Info. Theory*. **21**, 203–7.

Peterson, W. W. (1961). *Error-correcting codes*. MIT Press, Cambridge, Mass.

Peterson, W. W. and Weldon, E. J. (1972). *Error-correcting codes*, 2nd ed. MIT Press, Cambridge, Mass.

Pierce, J. R. (1973). The early days of information theory. *IEEE Trans. Info. Theory* **IT–19**, 3–8.

Pless, V. (1982). *Introduction to the theory of error-correcting codes*. Wiley, New York.

Plotkin, M. (1960). Binary codes with specified minimum distance. *IEEE Trans. Info. Theory* **6**, 445–50.

Plumstead, J. (1982). Inferring a sequence generated by a linear congruence. *IEEE FOCS* **23**, 153–9.

Pohlig, S. C. and Hellman, M. E. (1978). An improved algorithm for computing logarithms over $GF(p)$ and its cryptographic significance. *IEEE Trans. Info. Theory* **IT–24**, 106–10.

Pollard, J. M. (1974). Theorems on factorization and primality testing. *Proc. Camb. Philos. Soc.* **76**, 521–8.

Pollard, J. M. (1975). A Monte Carlo method for factorization. *BIT* **15**, 331–4.

Pollard, J. M. (1978). Monte Carlo methods for index computations (mod *p*). *Math. Comp.* **32**, 918–24.

Pomerance, C. (1983). Analysis and comparison of some integer factoring algorithms. In Lenstra, H. W. and Tijdeman, R. (eds.) *Number theory and computers*. North-Holland, Amsterdam.

Pratt, F. (1942). *Secret and urgent*. Blue Ribbon Books.

Purdy, G. B. (1974). A high security log-in procedure. *Comm. ACM* **17**(8), 442–5.

Rabin, M. O. (1976). Probabilistic algorithms. In Traub, J. F. (ed.), *Algorithms and complexity*, Academic Press, New York, pp. 21–39.

Rabin, M. O. (1978). Digitalised signatures. In de Millo, R. A. *et al.* (eds.) *Foundations of secure computing*. Academic Press, New York, pp. 155–68.

Rabin, M. O. (1979). Digitalized signatures and public key functions as intractable as factorization. *MIT Laboratory for Computer Science,* January, TR 212.

Rabin, M. O. (1980). Probabilistic algorithm for primality testing. *Journal of Number Theory* **12**, 128–38.

Rabin, M. O. (1980). Probabilistic algorithms in finite fields. *SIAM J. Comp.* **9**, 273–80.

Reeds, J. (1977). Entropy calculations and particular methods of cryptanalysis. *Cryptologia* **1**, 235–54.

Riesel, H. (1985). *Prime numbers and computer methods for factorisation.* Birkhauser.

Rivest, R. L. (1985). RSA chips (past/present/future). *Advances in cryptology: Proc. Crypto* **84**, 159–65.

Rivest, R. L. and Sherman, A. T. (1983). Randomized encryption techniques. In Chaum, D. *et al.* (eds.) *Advances in cryptology: Proc. Crypto* **82**, 145–63.

Rivest, R. L., Shamir, A., and Adleman, L. (1978). A method for obtaining digital signatures and public key cryptosystems. *Comm. ACM* **21**, 120–6.

Rosenberg, R. (1986). Slamming the door on data thieves. *Electronics,* February, 27–31.

Rumely, R. (1983). Recent advances in primality testing. *Amer. Math. Soc. (Notices)* **30**, 475–7.

Savage, J. E. (1976). *The complexity of computing.* Wiley, New York.

Schönhage, A. and Strassen, V. (1971). Schnelle Multiplikation grosser Zahlen. *Computing* **7**, 281–92.

Schroeder, M. R. (1983). *Number theory in science and communication.* Springer-Verlag.

Schwartz, J. T. (1979). Probabilistic algorithms for verification of polynomial identities. *Symbolic and Algebraic Computation,* Springer Lecture Notes in Computer Science **72**, 200–15.

Shamir, A. (1982). A polynomial time algorithm for breaking the basic Merkle–Hellman cryptosystem. *IEEE FOCS* **23**, 145–52.

Shamir, A. and Zippel, R. E. (1980). On the security of the Merkle–Hellman cryptographic scheme. *Trans. IEEE Info. Theory* **IT–26**(3), 339–40.

Shannon, C. E. (1948). A mathematical theory of communication. *Bell Syst. Tech. J.* **27**, 379–423 and 623–56.

Shannon, C. E. (1949). Communication theory of secrecy systems. *Bell. Syst. Tech. J.* **28**, 657–715.

Shannon, C. E. (1951). Prediction and entropy of printed English. *Bell Syst. Tech. J.* **30**, 50–64.

Shannon, C. E. (1957). Certain results in coding theory for noisy channels. *Information and Control* **1**(1), 6–25.

Shannon, C. E. and Weaver, W. (1949). *The mathematical theory of communication.* University of Illinois Press, Urbana, Illinois.

Simmons, G. J. (1979). Cryptology: the mathematics of secure communication. *Math. Intelligencer* **1**(4), 233–46.

Simmons, G. J. (1979). Symmetric and asymmetric encryption. *ACM Computing Surveys* **11**(4), 305–30.

Simmons, G. J. (1982). Message authentication without secrecy. In Simmons, G. J. (ed.) *Secure communications and asymmetric cryptosystems,* AAAS Selected Symposia Series, Westview Press, Boulder, pp. 105–39.

Simmons, G. J. (1983). Verification of treaty compliance—revisited. *Proc. IEEE 1983 Symp. on Security and Privacy,* 61–6.

Simmons, G. J. (1984). Message authentication: a game on hypergraphs. *Proc. 15th S. E. Conf. on Combinatorics,* Graph Theory and Computing, 161–92.

Simmons, G. J. (1985a). Authentication theory/coding theory. In Blakley, R. (ed.) *Advances in cryptology: Proc. Crypto* **84**, 411-31.

Simmons, G. J. (985b). Cryptology. *Encyclopaedia Britannica,* 16th edn., 913–24B.

Simmons, G. J. (1986). The practice of authentication. In Pichler, F. (ed.) *Advances in cryptology: Proc. Crypto* **85**, 261–71.

Singleton, R. C. (1964). Maximum distance q-nary codes. *IEEE Trans. Info. Theory* **10**, 116–18.

Sinkov, A. (1978). *Elementary cryptanalysis: a mathematical approach.* Random House, New York.

Slepian, D. (1973). Information theory in the fifties. *IEEE Trans. Info. Theory* **19**, 145-8.

Slepian, D. (ed.) (1974). *Key papers in the development of information theory.* IEEE Press, New York.

Sloane, N. J. A. (1981). Error-correcting codes and cryptography. In Klarner, D. A. *The mathematical gardner,* Wadsworth, Belmont, Calif., pp. 346–82.

Solovay, R. and Strassen, V. (1977). A fast Monte Carlo test for primality. *SIAM J. Comp.* **6**, 84–5. [erratum **7** (1978), 118].

Stephens, N. E. (1986). Lenstra's factorization method based on elliptic curves. *Advances in Cryptology: Proc. Crypto* **85**. Lecture Notes in Computer Science, 218. Springer-Verlag, pp. 409–16.

Strassen, V. (1969). Gaussian elimination is not optimal. *Numerische Mathematik* **13**, 354–6.

Thompson, T. M. (1983). *From error correcting codes through sphere packings to simple groups*. Math. Ass. of America.

Varshamov, R. R. (1957). Estimate of the number of signals in error correcting codes. *Dokl. Akad. Nauk SSSR* **117**, 739–41.

Vazirani, U. V. and Vazirani, V. V. (1984). Efficient and secure pseudo-random number generation. *IEEE FOCS* **25**, 458–63.

Vazirani, U. V. and Vazirani, V. V. (1984). RSA bits are $0.732 + \varepsilon$ secure. In Chaum, D. (ed.) *Advances in cryptology: Proc. Crypto* **83**, 369–75.

Vernam, G. S. (1926). Cipher printing telegraph systems for secret wire and radio telgraphic communications. *J. AIEE* **45**, 109–15.

Verriest, E. and Hellman, M. E. (1979). Convolutional encoding for Wyner's wiretrap channel. *IEEE Trans. Info. Theory* **IT–25**, 234–7.

Viterbi, A. and Omura, J. (1978). *Digital communication and coding*. McGraw Hill, New York.

Wagner, K. and Wechsung, G. (1986). *Computational complexity*. Reidel, Dordrecht.

Weaver, W. (1949). The mathematics of communication. *Scientific American,* July, 11–15.

Welsh, D. J. A. (1983). Randomised algorithms. *Discrete Appl. Math.* **5**, 133–45.

Western, A. and Miller, J. (1968). *Tables of indices and primitive roots*. Royal Soc. Math. Tables, Vol. 9. Cambridge University Press, London.

Wilkes, M. V. (1972). *Time sharing computer systems*. Elsevier, New York.

Williams, H. C. (1980). A modification of the RSA public key cryptosystem. *IEEE Trans. Info. Theory* **IT–26**, 726–9.

Williams, H. C. (1984). Factoring on a computer. *Math. Intelligencer* **6**, 29–36.

Williams, H. C. and Schmid, B. (1979). Some remarks concerning the MIT public-key cryptosystem. *BIT* **19**, 525–38.

Wolfowitz, J. (1961). *Coding theorems of information theory*. Prentice-Hall, New Jersey.

Wyner, A. D. (1975). The wire-tap channel. *Bell Syst. Tech. J.* **54**(8), 1355–87.

Yao, A. C. (1982). Theory and applications of trapdoor functions. *IEEE FOCS* **23**, 80–91.

Yavuz, D. (1974). Zipf's law and entropy. *IEEE Trans. Info. Theory* **20**, 650.

Zipf, G. K. (1935). *The psycho-biology of language*. Houghton Mifflin, Boston.

Zipf, G. K. (1949). *Human behaviour and the principle of least effort*. Addison-Wesley.

Index